A Level Product Design

Brian Evans

Will Potts

Published in 2004 by:
Nelson Thornes Ltd
Delta Place
27 Bath Road
CHELTENHAM
GL53 7TH
United Kingdom

04 05 06 07 08 / 10 9 8 7 6 5 4 3 2 1

A catalogue record for this book is available from the British Library

ISBN 0 7487 8674 0

Illustrations by Barking Dog Art except 'Stages involved in the use of GRP' on page 24 and 'slab sawing' and 'Different methods of quarter sawing' on page 61 by and copyright © Brian Evans 2004
Page make-up by Northern Phototypesetting Co Ltd, Bolton

Printed and bound in Croatia by Zrinski

Contents

Preface

One way of learning about the creation of products and problem solving is to look at products that already exist. We can look at a product and discover what materials have been used, then go on to think about why they have been used – were they chosen because of the manufacturing methods that are available or were they chosen with the user in mind?

For example, the manufacture of a shampoo bottle has to be cost-effective and so appropriate materials and processes should be chosen; whereas a piece of bespoke jewellery will use specialist materials chosen by the user, and the product will be produced by very skilled designers and craftspeople.

Products produced today also require very different techniques from those of only a short time ago: now techniques need to take account of environmental issues, for example how the materials are extracted and processed, the levels of energy used and whether or not the materials can be recycled.

The main aims of this book are two-fold:

● to provide a base of knowledge that reflects the requirements of study at AS and A2;

● to help develop your product analysis skills through the activities that are distributed throughout this book.

It is not intended that this book be read from cover to cover, neither is it intended to be a definitive reference work covering the complete requirements of all the examination boards. It has been written as a reference source that should support your school or college lessons, as and when required, while at the same time providing ample opportunities for product analysis and for revision.

This is a practical student book that will guide you through the main subject topics that you need to learn for AS and A2 Product Design. The book is divided into three sections:

● Materials and components;

● Processes and manufacture;

● Design and market influences.

Each section contains the basic information you need to know about each topic. Case studies are used to give you an insight into specific products and industries, and there are plenty of tasks and exam questions to help you learn each topic. Throughout the book you will find product analysis exercises that will help you learn about a wide range of materials, manufacturing processes and important design issues. There are also extra shorter tasks to test your understanding of each topic. The key words for each topic are explained in the glossary at the back of the book. A range of diagrams and photographs have been included to help your studies. A good number of the diagrams are drawn in a 'student style' and should give you an idea about the type and standard of any diagrams you might produce in your answers to examination-type questions. You may find this a useful revision aid.

You can use this book as a reference book, a revision guide and as a text to support your lessons. There is a CD-ROM accompanying this book that contains guidance on answering the product analysis exercises and examination-type questions. Throughout the book we have recommended further reading to support the material covered in the book. We strongly recommend that that you use a wide range of books, websites and CD-ROM resources in learning about Product Design.

We hope you find this book interesting and rewarding.

Brian Evans and Will Potts

Introduction

Product analysis exercises

Why should I analyse products?

Through analysing products you will build up an appreciation of the materials that are used in making similar products. You will start to question and understand why certain materials are used for certain product types, and how these materials are manipulated in the manufacture of products. By investigating how products are made, you will later be able to recognise how other similar products would be made and explain why specific manufacturing processes are used. You will also start to appreciate how the selection of specific materials and manufacturing processes influences other things such as:

- function;
- aesthetic styling features;
- ergonomic features;
- product life cycles;
- safety, etc.

 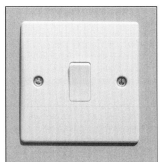  

This book has a series of product analysis exercises for you to complete. When analysing products you need to investigate:

- specific materials and how suitable they are for the product;
- specific manufacturing processes and the reasons for their use;
- how the choice of material and manufacturing process benefits the consumer and manufacturer;
- how the choice of material and manufacturing process affects the design of the product, e.g. styling/aesthetic features, ergonomics, etc;
- the selection of materials and processes and their impact on the environment;
- how the development of materials, technology and manufacturing processes has led to improvements in the quality and safety of products.

The product analysis exercises will ask you questions about each product. You can find the answers to the questions on the CD-ROM accompanying this book.

Product analysis exercise (a worked example)

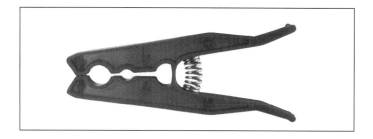

1. Name a suitable material for the clothes peg pictured above.

   High Impact Polystyrene (HIPS)

2. Explain why this material is suitable for the clothes peg

   - HIPS is a thermoplastic, which is suitable for injection moulding.

   - It is available in a range of colours, which make the product appealing to customers.

   - It is a durable material, which means it will withstand being left outdoors for some time.

   - It is rigid and hard, which help in the function of a clothes peg.

   - It is a relatively inexpensive material, as it is widely available. This is essential in a

     mass-produced product like a clothes peg.

   - The material is stable and the structure is consistent throughout.

     (Woods can have defects.)

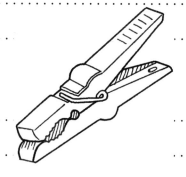

3. Explain why a clothes peg like this would be injection moulded.

   - The peg can be made in large volumes, quickly and

     continuously. This means that the unit cost (the cost

     per item) can be kept as low as possible.

   - Small details, such as textured grips, can be

     moulded without the need for further machining.

   - There is little waste compared to producing pegs

     from wood.

   - Multi-component moulding: lots of pegs can be

     moulded with every cycle by efficient mould design.

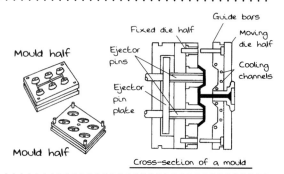

Mould half

Mould half

Guide bars

Fixed die half

Moving die half

Ejector pins

Cooling channels

Ejector pin plate

Cross-section of a mould

Using sketches in exams

In exams you may not be specifically asked to use sketches, but it is often very useful to use them in your answer to support what you are writing. For example, when trying to explain how ergonomic a mobile phone is, it would be a good idea to produce a labelled sketch showing the parts of the phone that are ergonomically designed, such as the shaping of the phone to fit a hand, the large, easy-to-read screen, and so on. Trying to explain this without a diagram would be difficult.

Step-by-step diagrams are also very useful when trying to explain how a product is made. Manufacturing processes, whether industrial such as injection moulding, or workshop-based such as turning, are difficult to describe without using diagrams, and very difficult for a reader to understand without seeing diagrams.

When using sketches, you should avoid spending too long on them. Keep them simple and clear. Avoid using red or green ink, as this is used by examiners, and don't waste time decorating diagrams. You will not normally gain marks by making diagrams pretty and you may run out of time later.

A2 exercises

You will be presented with a range of exercises that cover the three themes of: materials and components; processes and manufacture; design and market influences. Before you start the A2 exercises, we recommended that you review the AS theory and product analysis exercises in this book (and the accompanying CD-ROM, if you have access to it). A2 work builds on this: A2 students will be expected to develop a wider understanding of industrial practice and how design and market issues influence product design and manufacture.

A2 exercise (a worked example)

Traditional materials, such as metals and woods, are being replaced with polymers for the manufacture of a wide range of products. Using **two different** products, describe specific examples where traditional materials have been replaced with polymers and explain the benefits this brings to both the manufacturer and the consumer.

Product 1: Window frames

Window frames have traditionally been manufactured from softwood, such as pine. Pine is a suitable material for window frames as it is in plentiful supply and therefore relatively inexpensive, and it can be easily machined to make the rebates for the glazing, etc. It can be stained, treated with preservative or painted to the customer's requirements.

However, there are disadvantages in making window frames from timber: the length of time it takes to make the frame (usually quite a skilled job); the on-going maintenance associated with timber frames, because timber will rot if unprotected; and the fact that timber expands and contracts with changes in temperature and humidity, which can affect the function of the windows.

Timber window frames are, gradually, becoming superseded by uPVC (rigid PolyVinylChloride) frames. This is because uPVC has many advantages over traditional timber alternatives. Firstly, the uPVC frames are made up of extruded profiles, which can be manufactured almost continuously using modern plastic extruders. Unlike timber frames, there is very little waste in manufacture because precise quantities of plastic are used and there is no need to machine the plastic. Plastic welding easily joins uPVC, whereas timber frames normally need to be joined using joints such as mortice and tenon. uPVC window frames do not need any maintenance other than occasional cleaning; they do not need painting, etc. uPVC is a stable material, in that it will not warp or swell with changes in temperature and humidity, and it is not affected by sunlight.

Although uPVC is more expensive than softwood, the fact that little maintenance is required and that uPVC is very long lasting outweighs the costs.

Product 2: Drinks bottles

Drinks bottles have traditionally been made from glass. Glass is used because it can be moulded. This enables aesthetic styling features and ergonomic features, such as the shaping of the bottle to fit the hand or textured sections for grip, to be moulded in relatively easily. Glass is available in a variety of colours and, of course, is transparent to enable the contents to be seen. Glass is food safe and it can be recycled.

The main disadvantages in using glass for the packaging of drinks bottles are that glass requires a considerable amount of energy to make into a molten state for moulding and for recycling. In addition to this, discarded broken glass can present dangerous environmental hazards. This also has some implications for distributors and retailers who need to take care in handling and storing glass containers.

For the reasons outlined above, glass bottles have gradually been replaced by plastic ones. Typically, plastics such as Polyethylene Terephthate (PET) and High Density Polyethylene (HDPE) are used to manufacture bottles.

Manufacturing bottles from plastics has several advantages: the bottles can be blow moulded at a much lower temperature than glass, thus saving energy costs; plastics are much more durable than glass and therefore can be transported and displayed more safely. This property also makes it possible to retail bottles in vending machines. As plastics can be moulded (like glass), the required aesthetic and ergonomic features are relatively easy to manufacture. PET and HDPE are food safe and, as they are thermoplastic, they can be recycled. Recycling is an increasingly important aspect, as manufacturers are becoming legally obliged to use recycled or recyclable materials.

Materials and components

Introduction to Part One

As a student of product design it is important that you become familiar with the range of materials available. You should be able to identify materials and give reasons why they have been used at AS level, while at A2 you should be sufficiently familiar with the characteristics of materials to be able to make informed decisions as to the suitability of specific materials for particular applications.

This first part of the book looks at the main groups of materials, i.e. plastics, metals and woods, as well as ceramics and glass. Composite materials are also discussed along with the newer SMART materials. In addition you will find reference to some manufacturing processes in this section where appropriate.

Reference has also been made to materials' sources. This has been treated as an overview only and is meant as background information when considering appropriate use of materials.

You will need to make use of other resources – books, CD-ROMS, the Internet, and so forth – to find all the information you will need for success in product design. Some references for further reading are given in this section.

Plastics

Introduction to plastics

Plastics (or more correctly polymers) play an increasingly important part in the design and manufacture of everyday products. Designers today need to be aware of the range of plastics available and how plastics can be processed in order to produce successful products.

At AS level
As an AS level student, you also need to know about the range of plastics available along with their general properties and characteristics. You should also be familiar with the more popular methods of manufacturing and the role of a variety of additives in the successful manufacture and use of products.

At A2 level
A2 level students should build on what they have learned at AS level in order to provide a reasoned response to examination questions, for example why plastics have successfully replaced traditional materials in the manufacture of specific products. See 'Advantages of replacing glass with polymers', p.71.

What are plastics?

Plastics are a group of materials made up of long chains of molecules. A large proportion of them are known as 'hydrocarbons', because the long-chain molecules consist of hydrogen and carbon atoms in various combinations with other atoms.

Why are plastics so popular?

Polypropylene jug kettle

Plastics are becoming more popular in the manufacture of products, and in some cases have replaced more traditional materials. For example, the domestic kettle has developed from a product manufactured from a metal (for example, stainless steel) to a polymer with all the inherent benefits this material brings (lightweight, self-coloured and self-finished, electrical and a thermal insulator). In addition, the processes used in the manufacture of the polymer kettle have enabled the product to become more 'user friendly', by incorporating all of the ergonomic issues associated with the use of the product.

Sources of plastics

There are a number of sources of raw material used in the production of plastics (polymers).

- Animal and vegetable by-products are used in the manufacture of Semi-Synthetic Polymers. For example, cellulose is a natural material occurring in plant fibres. This is mixed with acetic acid to produce cellulose acetate from which products such as OHP slides are manufactured.

  Casein is a further example. In this case, by-products of milk are used to create the polymer. The material is still used in the manufacture of buttons.

- Coal, oil and gas are the sources for a range of purely synthetic plastics. Plastics produced from these materials are gained through the process of 'thermal cracking'. Synthetic plastics are therefore carbon-based and account for the majority of plastics used today.

Types of plastic

Plastics can be grouped into three types.

- **Thermoplastics:** these materials can be repeatedly reheated and remoulded.
- **Thermosets (or thermosetting plastics):** these undergo a chemical change resulting in them becoming permanently rigid, i.e. they cannot be reheated and reshaped.
- **Elastomers:** these are polymers that have good elasticity, i.e. they can be distorted under pressure but will return to their original shape when the pressure is removed.

The diagram below shows examples of the different types of polymer available for use in manufacturing products; more details of common polymers are given in Table 1 (overleaf).

The main types of plastics (polymers)

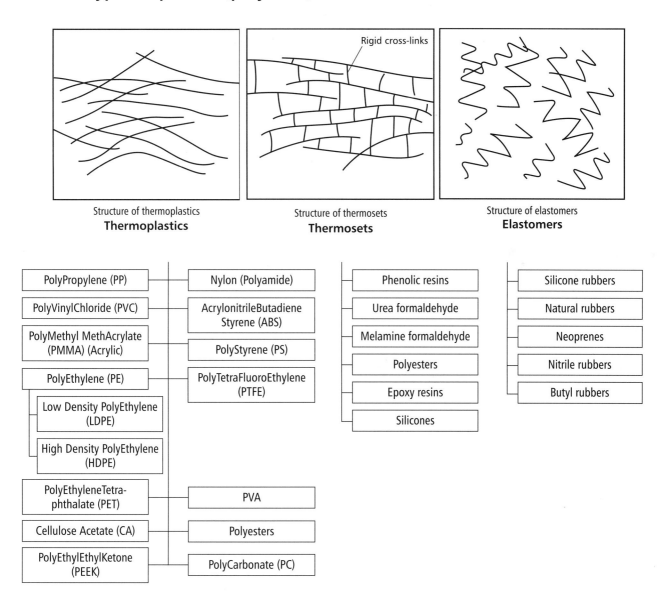

Structure of thermoplastics
Thermoplastics

Structure of thermosets
Thermosets

Structure of elastomers
Elastomers

Common products made from plastics

Table 1: Common polymers

Common name	Working name	Characteristics	Common uses
THERMOPLASTICS			
ABS	Acrylonitrile Butadiene Styrene	High impact strength, giving good toughness with good strength, scratch-resistant, lightweight and durable	Kitchen products, mobile telephone cases, PC monitor cases, safety helmets, toys, some car parts and domestic telephones
CA	Cellulose Acetate	Tough and rigid, lightweight with good strength, transparent and non-flammable	Photographic film, handles for cutlery, cupboard door knobs, frames for glasses
Nylon	Polyamide	Hard, tough, resistant to wear with a low coefficient of friction	Bearings, gears, curtain rail fittings and clothing
PMMA (PolyMethyl MethAcrylate)	Acrylic	Food-safe, tough, hard, durable, easily machined	Light units, illuminated signs, lenses for car lights
PP	PolyPropylene	Lightweight, food-safe, good impact resistance even at low temperatures, good chemical resistance	Kitchen products (food containers), medical equipment, string and rope
HIPS	High Impact Polystyrene	Good impact resistance, good strength and stiffness, lightweight	Toys and refrigerator linings
PS	Polystyrene	Lightweight, rigid, colourless, low impact strength	Packaging, disposable cups/plates and containers
	Expanded polystyrene	Floats, good sound and heat insulator, lightweight, low strength	Packaging, disposable cups, sound and heat insulation
LDPE	Low Density PolyEthylene	Low density (lightweight), low stiffness and rigidity, good chemical resistance	Detergent bottles, toys and carrier bags
HDPE	High Density PolyEthylene	High density, good stiffness, good chemical resistance	Crates, bottles, buckets and bowls
uPVC	PolyVinyl Chloride	Good chemical resistance, good resistance to weathering, rigid, hard, tough, lightweight, can be coloured	Pipes, guttering, bottles and window frames

PVC (un-plasticised, flexible)	PolyVinyl Chloride	Good chemical resistance, good resistance to weathering, rigid, hard, tough, lightweight, can be coloured	Flexible hose, e.g. hose pipes, cable insulation
PET	PolyEthylene Terephthalate	Moderate chemical resistance	Fibres used to make a wide range of clothing, blow-moulded bottles for beers and soft drinks, electrical plugs and sockets, audio and video tapes, insulation tapes
PC	PolyCarbonate	Good chemical resistance, expensive material	Very tough – used for protective shields, e.g. safety glasses, safety helmets, hairdryer bodies, telephone parts, vandal-proof street-light covers
THERMOSETS			
Epoxy resins		High strength when reinforced with fibres (GRP – glass-reinforced plastic), good chemical- and wear-resistance	Surface coating, encapsulation of electronic components, adhesives
Melamine formaldehyde		Rigid, good strength and hardness, scratch-resistant, can be coloured	Tableware, decorative laminates for work surfaces
Polyester resins		Rigid, brittle, good heat and chemical resistance	Casting, used in GRP (e.g. boat hulls and car body parts)
Urea formaldehyde		Rigid, hard, good strength, brittle, heat-resistant, good electrical insulator	Electrical fittings, adhesives

General properties of plastics

Here is a summary of the properties of plastics.
- They are good electrical and thermal insulators.
- They have a good strength to weight ratio – this does not mean they are strong materials in the same way that mild steel is strong, but they have good strength compared to their weight.
- Generally, they have good atmospheric and chemical corrosion resistance.
- They have fairly low melting temperatures, ranging from 70 to 185 °C (thermoplastics only).
- They are lightweight.
- They can be self-coloured, opaque, translucent or transparent, depending on the type of polymer and processing used.

Improving the properties of plastics: additives

A variety of materials can be incorporated into the polymer powder prior to processing. Some of these, e.g. fillers, can be used to give the material bulk,

while others are used to condition the material (i.e. increase the mechanical properties of the material), in a similar way to heat treatments for metals. For example, particular additives give the material anti-static properties, make the material flame retardant or resistant to ultraviolet light.

Fillers

Fillers are used to reduce the bulk of the plastic. They are generally cheaper than plastics and so help reduce costs. Examples of fillers include: sawdust, wood flour, crushed quartz and limestone. Some fillers can increase strength and hardness of the polymer by removing brittleness.

Flame-retardants

Flame-retardants are used to reduce the risk of combustion. Their main role is to create a chemical reaction once combustion has begun; they release agents that will stifle the combustion. An example of the use of flame-retardants is in the foams used to fill seating cushions.

Anti-static agents

Anti-static agents reduce the effects of static charges that could build up on a product, e.g. from walking on a carpet made from synthetic materials.

Plasticisers

Plasticisers are added to plastics to improve the flow properties of plastics when being moulded. They also reduce the softening temperature and go some way to making the material less brittle.

Stabilisers

Stabilisers are used to reduce the effects of ultraviolet light, i.e. by making the plastics more resistant to being 'broken down' by sunlight. This is important both from a structural and an aesthetic point of view. Stabilisers are used in products that are exposed to a lot of sunlight (such as windows, doors and conservatory components).

Applications for thermoplastics

There is a huge range of thermoplastics available. Further development of these materials extends their usefulness. Examples of improved polymers include those that have been designed to meet food quality standards. The development of polymers like PET helps to extend the shelf life of carbonated drinks; these polymers are less permeable than previously used polymers such as PVC.

Being thermoplastic, these materials will begin to soften at raised temperatures. Again further development has helped produce materials that will withstand higher temperatures than previously used polymers, allowing motor casings to be moulded integrally. The ability to withstand higher temperatures has enabled the manufacture and use of products such as 'ready-meals', where the plastic container is placed in the oven with the food as it cooks – something that was unheard of a few years ago.

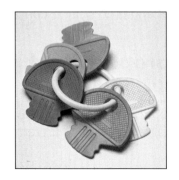

Thermoplastic products

Processing plastics

Thermoplastics can be processed in a number of ways depending on the shape of the product being manufactured, while thermosets are limited in the manufacturing methods used. The diagram below gives an overview of processes used.

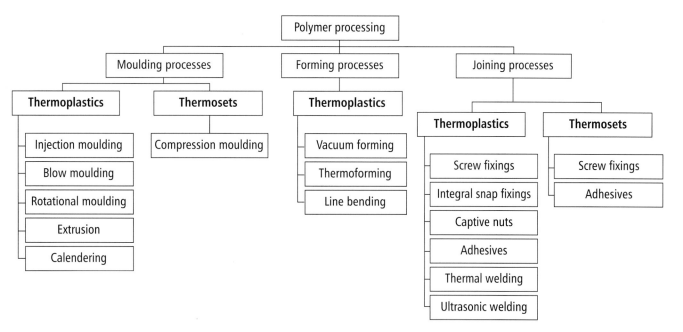

We will now consider some of the more common moulding and forming processes for thermoplastics. Joining processes will not be covered here. For more on joining for polymers see p.98.

Further reading
- *Materials for the Engineering Technician, Third Edition*, R A Higgins (Arnold)
- *A Level Design and Technology, Third Edition*, Norman, Cubitt, Urry and Whittaker (Longman)

Injection moulding

This process is most commonly associated with thermoplastics and is used to produce complex three-dimensional shapes.

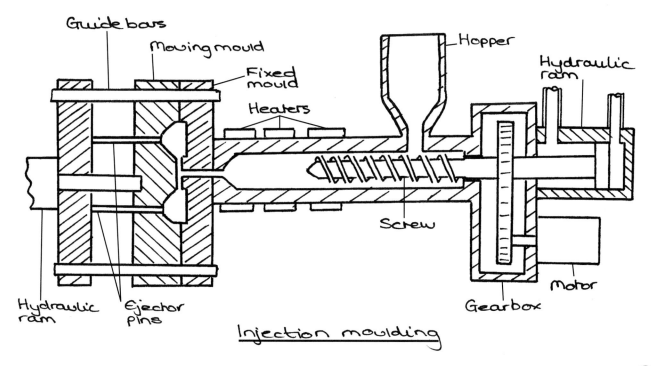

Injection moulding

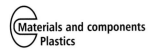

Stages of the process:

Step 1 Plastic granules (plus any other additives and colours mixed with them) are placed in the hopper. The granule mixture falls through the hopper onto the Archimedean screw.

Step 2 The screw is rotated via the motor and gearbox. This action forces the polymer forwards towards the heaters, where it becomes softened to the point where it is ready to be injected into the mould.

Step 3 The hydraulic ram forces the softened polymer through the feedhole into the mould. Pressure from the ram ensures the mould cavity has been filled.

Step 4 When sufficient time has passed to allow the polymer to cool and solidify (a matter of seconds), the mould halves are opened. As they open, ejector pins are activated to release the product from the mould.

Step 5 Once emptied, the mould is then closed ready to begin another cycle.

Advantages and disadvantages of injection moulding

Advantages
- Very complex three-dimensional shapes can be produced.
- High volumes can be produced with consistent quality.
- Metal inserts can be included in the item being produced.

Disadvantages
- Initial set-up costs are high.
- Moulds are expensive.

Task

Identify three products produced by injection moulding.
Identify the signs that suggest a product has been injection moulded.
State the materials used to manufacture the products.

Blow moulding

This process is used in the manufacture of bottles and other containers. Objects produced are usually hollow and have a narrow neck.

Stages of the process:

Step 1 A tube of heated and softened polymer is extruded vertically downwards. This tube is called a Parison.

Step 2 The mould halves close trapping the upper end of the Parison, effectively sealing it.

Step 3 Hot air is then blown into the Parison forcing it out to follow the shape of the mould.

Step 4 The mould effectively cools the polymer allowing it to be released from the mould.

Step 5 The mould halves are opened and the product is extracted.

Advantages and disadvantages of blow moulding

Advantages of blow moulding
- Once set up, blow moulding is a rapid method of producing hollow objects with narrow necks.
- Non-circular shapes can be produced.

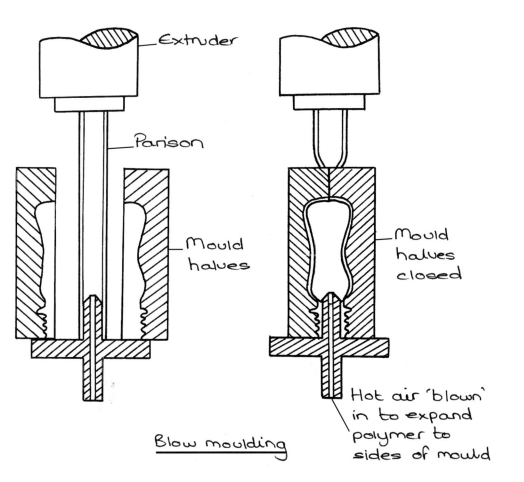

Extruder

Parison

Mould halves

Mould halves closed

Hot air 'blown' in to expand polymer to sides of mould

Blow moulding

Disadvantages of blow moulding
- Moulds can be expensive.
- It's difficult to produce re-entrant shapes, i.e. shapes that do not allow easy extraction from the mould (e.g. a dovetail joint).
- Triangular-shaped bottles are difficult to produce.

Task
Identify three different products that have been manufactured by blow moulding. Identify the materials used in these products.
State the main characteristics of products that are produced in this way.

Blow moulded products and mould

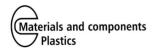
Rotational moulding

Rotational moulding is used in the manufacture of three-dimensional hollow products, such as footballs, road cones and large storage tanks (up to 3 m³ capacity).

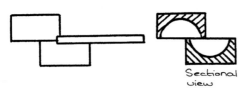

1. Open mould is filled with plastic powder.

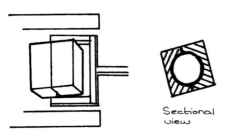

2. Mould is heated and the plastic melts, coating the inside.

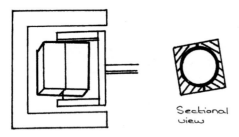

3. Mould is cooled to set the plastic.

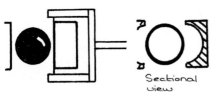

4. Mould is opened and the product removed.

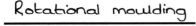

Rotational moulding

Stages of the process:

The machines used have a number of arms that rotate about a fixed central point. Moulds are attached to the end of each arm and are rotated continuously. The only time the moulds do not rotate is when they are at the starting point and end point of the process.

Step 1 Once the moulds have been loaded with a precise weight of thermoplastic powder (e.g. polyethylene) the mould halves are clamped together.

Step 2 The moulds then are rotated about the arm spindle and the whole arm is rotated towards a heated chamber where the thermoplastic material is heated to its melting point. The continuously rotating mould ensures that the thermoplastic covers all of the mould.

Step 3 The next stage of the process is the cooling chamber where the material is cooled ready to be extracted from the mould.

Step 4 The mould is then returned to the starting point where the mould halves are separated and the product removed.

Cycle times vary as they depend on the required wall thickness of the material.

Advantages and disadvantages of rotational moulding

Advantages

- One-piece mouldings can be produced.
- It is ideal for both rigid, tough shapes and flexible shapes.
- A large range of sizes is possible, from small medical components to large storage tanks.
- Surface textures can be applied to the finished products from textures applied in the mould.
- Moulds tend to be cheaper than those for injection or blow moulding, since high pressures are not required.
- Cheaper moulds allow lower production runs.

Disadvantages

- Only hollow shapes can be produced in this way. More complex three-dimensional shapes would either be blow moulded or injection moulded.

Thermoforming and vacuum forming

Thermoforming is a relatively new process, but is very closely related to vacuum forming. Where vacuum forming relies solely on a vacuum to 'pull' the softened polymer around a mould, thermoforming uses an outer mould to help in the process – this allows a greater level of detail, such as lettering, symbols and sharp edges, to be achieved.

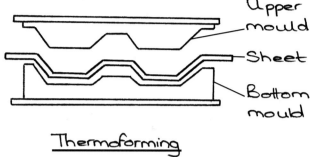

Thermoforming

The thermoforming process

Sheet material is heated to just above its softening point and then held securely in a frame between the two mould halves. The mould halves close and at the same time a vacuum is applied through the lower mould. The upper mould ensures the required amount of detail is achieved.

Advantages and disadvantages of thermoforming

Advantages
- It's a low-cost process.
- It's good for smooth shapes with additional detail.

Disadvantages
- Deep moulds result in a thinning of the wall thickness where it has been stretched.
- It's limited to simple designs.
- Trimming is usually needed.

Task

Identify three further products that are thermoformed. State why thermoforming is appropriate for these products.

Extrusion

Extrusion is the process used where products with a continuous cross-section are required.

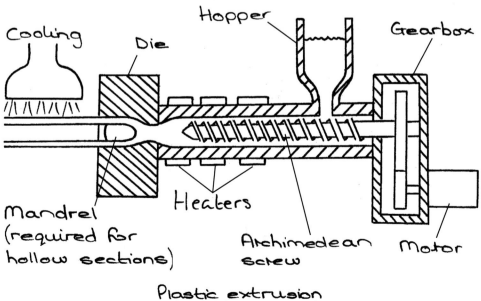

Plastic extrusion

In essence, the process forces molten plastic through a die that has the required cross-sectional shape.

Stages of the process:

Step 1 Thermoplastic powder is placed in the hopper; this powder then falls onto the rotating Archimedean screw, which in turn pushes the material towards a heated section of the extruder.

Step 2 The heaters soften the plastic, which is then forced through the die by the rotating screw.

Step 3 On exiting the die, the plastic product is then cooled using a water jet.

Step 4 Further along the transfer table, the product is cut to the required length.

Wires can be insulated with the aid of a special mandrel arrangement that allows the wire to pass through.

Advantages and disadvantages of extrusion

- Extrusion has the advantage of generally being a low-cost process that requires only simple dies.
- Its main disadvantage is that it can only produce continuous cross-sectional shapes.

Task

Identify three products produced by extrusion. Sketch the cross-section of each of the products identified. State the material used for each product.

Calendering

Calendering is used for the manufacture of thermoplastic film, sheet and coating materials. In the main, materials such as PE, PVC, ABS and cellulose acetate would be processed in this way. The LDPE bag on p.6 would have been calendered.

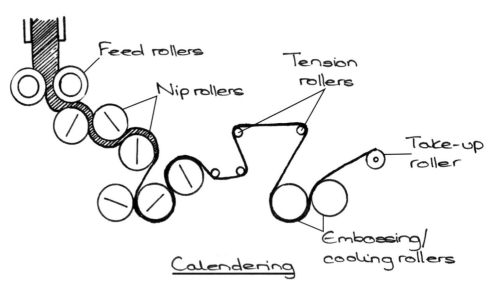

Calendering

Stages of the process:

Calendering involves rolling out a mass of pre-mixed plastic material between large rollers to form a continuous film of accurate thickness.

Step 1 The rollers are heated to just above the softening point of the thermoplastic.

Step 2 During the rolling process, the plastic 'dough' is forced through the gap roller. These rollers determine the thickness of the material.

Step 3 The final roller is the 'chill' roller that cools the material.

Task

Identify three products produced by the calendering process. State the materials used.

Line bending

Line or strip bending is used to form straight, small curved bends in thermoplastic sheet material.

The equipment used in this type of process is known as a 'strip heater' and comprises an electrical heater – usually a tensioned resistance wire – that is enclosed in a channel in a table. The sheet of thermoplastic material is clamped accurately over the heating element, so that the heater softens that part of the sheet that requires bending. In some line bending arrangements heating elements can be placed on both sides of the sheet, avoiding the need to turn the material over.

Once heated to the required temperature, the material can be bent. Accurate bends are achieved using bending jigs that ensure the correct angle is consistently being achieved.

Example of product produced by line bending

Applications for thermosetting polymers

Unlike thermoplastics, thermosetting plastics display very little plasticity. In other words, they are very rigid materials. There are fewer thermosetting plastics than there are thermoplastic materials, but they have useful properties that allow them to be used in a range of useful applications (see p.5).

For example, urea formaldehyde is a very stiff plastic that is hard and has good strength. In addition, in common with most plastics, it has very good electrical resistance properties (a good insulator) and, because it will not be softened by heat, it is a suitable material for electrical fittings.

Other applications for thermoset plastics include the use of polyester resin for paperweights. Polyesters can also be used in combination with a catalyst and glass fibres to create GRP – glass-reinforced plastic.

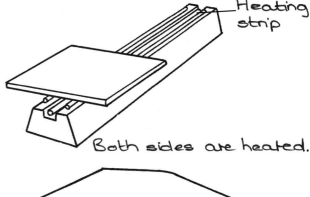

Heating strip

Both sides are heated.

When soft enough the sheet is bent around a bending former (or jig)

Line bending

Compression moulding

Compression moulding is probably the most important moulding process for manufacturing with thermosetting plastics. A combination of heat, pressure and time is needed to ensure all of the material's form and structure changes.

Stages of the process:

Step 1   A preformed 'slug' (compressed powder) of material is placed between the two halves of the mould.

Step 2   The mould is heated to a temperature that will allow the cross-links to form within the material.

Step 3   The mould is closed onto the preform and the pressure used will force out any excess material. The moulds are held closed under pressure at the required temperature for a period of time that is sufficient to allow all of the material to be 'cured' – all cross-links formed.

Step 4   When the mould is opened, the product can be ejected while it is still hot – it does not have to be cooled – and the process can begin again.

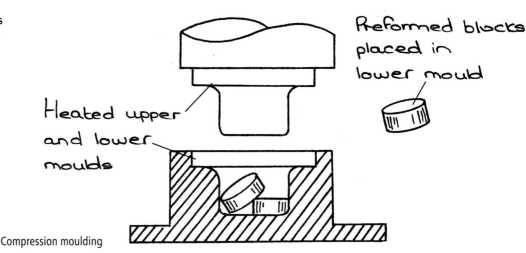

Heated upper
and lower
moulds

Reformed blocks
placed in
lower mould

Compression moulding

Advantages and disadvantages of compression moulding

Advantages

- Moderately complex parts can be produced over long production runs.
- Although there is some heavy machinery involved, start-up costs are relatively low – moulds are less expensive than those used in injection moulding.
- There is little waste material.

Disadvantages

- It is necessary to manufacture a preform.
- The process is restricted to products with generally low complexity.

> **Task**
> Give three applications where thermoplastics have been used.
> Identify the type of thermosetting plastic used and the main characteristics of the plastic that make it appropriate for the application.

Thermosetting and thermoplastic materials

This is a brief note highlighting those materials that behave in a similar way to polymers in that they demonstrate thermoplastic, thermosetting or elastomeric properties.

Plastics have a particular structure. The following diagrams show the structure of the three main types.

Thermoplastics

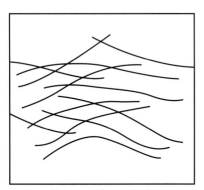

Structure of thermoplastics

Thermoplastics can be considered as a tangle of long-chain molecules. These molecules are held together by strong electrostatic forces, called van der Waals bonds. These bonds can be released with the application of heat, making it possible for the material to be reshaped.

Thermosets

Thermosetting plastics also consist of long-chain molecules but differ in that these molecules are held together by rigid cross-links, which in turn prevent them from being reheated and reshaped.

Elastomers

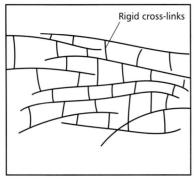

Rigid cross-links

Structure of thermosets

Elastomers have long-chain molecules that can be considered to be coils – a bit like springs – so that when the material is distorted (compressed or stretched) the molecules distort, and when released they return to their original shape. These coiled molecules give the elastomers their elastic properties.

At rest

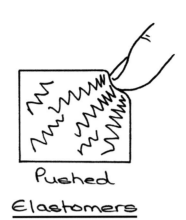

Pushed

Elastomers

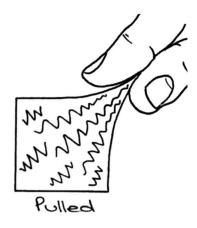

Pulled

Sketch showing the flexibility of elastomers

Other thermoplastic and thermosetting materials

Thermoplastic materials

Thermoplastic materials, once shaped, can be reshaped following the application of (i) heat or (ii) water.

For example, a thermoplastic polymer can be reshaped by heating; heating releases the van der Waals bonds that effectively link the long-chain molecules, allowing the material to be formed into a different shape. (Whereas a thermosetting polymer has rigid cross-links, which cannot be released by heating and therefore the material cannot be reshaped.)

- Clay is a good example of a thermoplastic material. If it is still in its natural state (i.e. has not been fired) then it can be softened and reshaped when mixed with water. However, when clay has been fired, it becomes a much more rigid material that behaves as a thermosetting material (i.e. rigid cross-linking has occurred and so it is not affected by water).

- Paper is another example of a material that can be broken down when mixed with water. This is the basis of paper recycling.

- Metals are another group of materials that behave in a similar way to thermoplastics when heated. Because of their structure, metals can be softened through the application of heat, making reshaping easier. Indeed, most metals can be heated until they become liquid and can be cast into shape.

- Glass is a further example of a material that can be heated and reshaped, behaving in a similar way to a thermoplastic.

Thermosetting materials

As described above, there are a number of materials that once processed by heat or a chemical reaction cannot be returned to their basic components and reshaped. The most common (non-polymer) example is ceramic materials. This group of materials includes:

- cements, concretes, refractory materials, for example, firebricks that may be used in applications such as domestic (gas) fires and industrial kilns;

- house and engineering bricks;

- products such as plant pots, paving slabs and wall tiles.

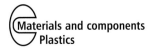

Table 2: Examples of non-polymer thermoplastic and thermosetting materials

Thermoplastic materials	Thermosetting materials
Clay, in the 'wet' state only – used for model making, ceramic figures and sculptures	Clay, once fired at a high temperature – used for plant pots, domestic crockery, wall and floor tiles
Paper – can be reformed once mixed with water (and some new fibres) to make recycled paper	Concrete – once the chemical reaction between the constituent parts has taken place and the product has dried out, then it cannot be remixed/reshaped
Metals – a good example is the steel used in the manufacture of cars: once separated from other materials it can be heated to the melting point of steel and reshaped using one of the many processes available for metals	
Glass – though very brittle materials, glasses can be heated to their melting point and reshaped	

Identifying plastics' processes

On close inspection of a product, it is possible to establish the method used to produce it.

Injection moulding

Those products that have been injection moulded usually have a complex three-dimensional shape that can only be produced by a moulding technique. An example of this would be the components of a computer mouse. The cover of the mouse is a very complex shape with curves and fixings for other components. In addition there are some ejector pin marks, another indicator that injection moulding has been used to produce this component.

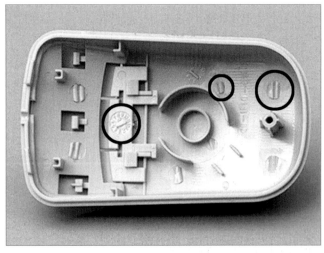

Photo showing a component made by injection moulding

Blow moulding

Products that are blow moulded have hollow shapes with one end sealed, while the other end remains open. This will include products such as drums for holding chemicals and liquids, bottles for spring water, and so on.

In blow moulding the 'neck' of the product is usually smaller than the outside measurement of the main body.

Surface decoration can be applied. For example, raised (embossed) lettering, a product logo or a recycling symbol. Screw threads can also be incorporated into the neck of the product to secure the lid.

Other signs of blow moulding include the sprue produced, usually at the bottom of the product, and mould seam lines around the outside of the product showing where the mould halves have been joined together.

Rotational moulding

All rotationally moulded products are hollow. There are no sprues, but there will be seam lines where the mould halves have been joined together.

As in blow moulding and injection moulding, embossing can also be included.

Thermoforming

Products that are thermoformed (and for that matter vacuum formed) are made from sheet materials. An amount of stretching occurs in both of these processes, making the sides of the product thinner than the base. There are no sprue marks, but there will be marks showing where the product has been cut and removed from the sheet.

In addition, thermoformed products contain sharper detail – such as lettering – than vacuum-formed products.

International symbols

Look at any modern plastic product and somewhere on it will be a symbol that identifies the type of plastic material that has been used in its manufacture – see Table 3.

Table 3: International symbols for polymers

SPI code	Type of polymer	Common uses
1 PETE	PETE (or PET), polyethylene terephthalate	Soft drinks and water bottles; deli and baking trays; oven-safe film and food trays; carpets and fibre filling
2 HDPE	HDPE, high density polyethylene	Milk, juice, shampoo, butter and yoghurt containers; grocery, rubbish and retail bags; cereal box liners; heavy-duty pipe; bottles for laundry products, oil and car washing fluid
3 V	V (or PVC), polyvinyl chloride	Pipes, film, clear packaging and carpet backing; containers for non-food items
4 LDPE	LDPE, low density polyethylene	Bread, frozen food and dry cleaning bags; carrier bags; squeezable bottles
5 PP	PP, polypropylene	Yoghurt and margarine containers; medicine bottles; car parts, carpets, industrial fibres
6 PS	PS, polystyrene	Meat trays, cups, plates, cutlery and compact disc sleeves; video- and audio-cassette cases
7 OTHER	Other (any polymer or combination of polymers not covered by categories 1-6), includes ABS, acrylonitrile butadiene styrene	Reusable water bottles, trays for the microwave; mobile telephone outer cases, computer parts, monitors, keyboard parts

A2: polymers

At A2, you would be expected to have a good level of knowledge of a wide variety of plastics and be able to understand how these plastics can be applied to a wide range of products.

By the time you take your exams, you should be confident to identify specific plastics used in everyday products. You should be able to explain how they are manufactured, and why these methods are used. The use of plastics in product design and manufacture brings many benefits to the manufacturer and the consumer. You should be able to understand and recall these benefits.

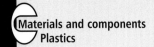

Case study: Plastics

Plastics used in detergent packaging

Bottled detergents

With the invention of liquid washing detergents some time ago, bottle packaging was required. The type of bottle shown in the photograph is commonly used in the storage and retail of such liquid detergents.

It is made from a tough plastic, such as polypropylene, with the top being made from high density polyethylene. Detergent bottles like these are made from durable plastics, so that they can be kept by the consumer for some time and re-filled using the sachet-style packaging shown below. As HDPE and PP are thermoplastics, it is possible to recycle the bottles. They can also be made from a percentage of recycled plastics. This is an important consideration for manufacturers, as they are required by legislation to meet ecological targets.

The main body

The main body of the bottle is made using blow moulding. This is a relatively fast process that is important in the manufacture of a mass-produced bottle. It is possible to mould in ergonomic features such as a comfortable handle used to carry the bottle and to pour the liquid.

The bottle top

The bottle top and pouring spout are injection moulded. This process uses molten plastic injected at high pressure into precision-made dies. It enables the manufacture of accurate components such as the bottle top, complete with textured grip to help in undoing and tightening the lid and an internal thread for joining the top to the bottle.

Advantages of using plastics

- The durability of HDPE and PP has obvious benefits to both the manufacturer and the consumer. The main benefit is that the bottles are less likely to burst, either in handling during warehousing and transport or in transit from the shops to home. This is an important functional aspect of bottle packaging.
- As plastics are 'self coloured' by adding a pigment to the polymer, the packaging can be coloured to meet aesthetic requirements. This is very important as colour is a key feature to brand identity.

Disadvantages of using plastics

- If the consumer does not re-fill the bottle, the plastics used would be wasted. If the consumer does not recycle the bottle, it would have to go to incineration or land-fill. Polypropylene and polyethylene are not biodegradable; therefore landfill is not a particularly good option.
- The bottle shape is awkward to store, transport and display. This is why this type of bottle is often made with flat sides to minimise such problems.

Sachet detergent packaging

This type of packaging was developed shortly after the bottle as a method of re-filling bottles. The sachet is retailed at a slightly lower price than a bottle, therefore encouraging consumers to buy these instead of new bottles each time they need more detergents.

Refill packaging

The sachets are made from a low density polyethylene film, using the calendering process. The graphics, branding etc. are applied to the film using offset lithographic printing methods, prior to the polymer being cut and plastic welded into the bag form.

In addition to the sachets being less expensive than the bottle, they are generally much easier to transport. This is because they are lighter and, perhaps more importantly, can flex into gaps, tessellating together inside cardboard boxes or loose on display.

As packaging plays such an important part in the transport and retail of products such as detergents, manufacturers are constantly aiming to develop new and exciting packaging methods. These can be strongly influenced by changes in consumer tastes or fashion, environmental pressures or developments in technology. Some of these influences will be examined later in this book.

Task

Think of another form of packaging that uses plastics and list the benefits that the use of plastics brings. Some key words you might want to consider when doing this are:

- function;
- performance;
- ergonomics;
- quality;
- cost;
- safety.

Further reading

This has been just a brief look at polymers and their uses. There is a vast range of these materials and within each major type (e.g. polystyrenes) there are hundreds, if not thousands, of variations. To cover them all would require a huge book and is certainly beyond the scope of this one. If you do wish however to delve deeper into plastics then there are a range of publications available to help you, including:

- *Materials for the Engineering Technician, Third Edition*, R A Higgins (Arnold)
- *A Level Design and Technology, Third Edition*, Norman, Cubitt, Urry and Whittaker (Longman)

PRODUCT ANALYSIS EXERCISE: *thermoplastics*

Plastic drinks bottle

1. The drinks bottle in the photograph is made from thermoplastics. Explain what is meant by the term 'thermoplastic'.

2. Name a suitable plastic for the bottle.

3. Explain why this material is suitable.

4. The bottle has been made using blow moulding. Use notes and diagrams to explain this process.

5. Name a suitable material for the bottle top.

6. Explain why this material is suitable.

7. Explain how the designer has considered the environment in the development of this bottle.

PRODUCT ANALYSIS EXERCISE: *thermosets*

Domestic light switch

1. The light switch in the photograph is made from a thermosetting plastic (thermoset). Explain what is meant by the term 'thermosetting plastic'.

2. Name a suitable thermosetting plastic for the light switch and explain why it is suitable.

3. Use notes and diagrams to show how the light switch could have been manufactured.

4. Explain why the light switch has been manufactured using this method.

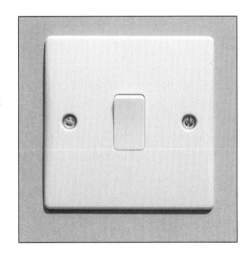

Task
Products such as electrical plugs, saucepan handles, picnic plates and cups are also made from thermosetting plastics. Use the Internet to find out which specific plastics are used and why.

Exam questions

AS exam question

1. Study the photograph above and answer the following questions:

   (a) Name a specific polymer suitable for the manufacture of
       the chair.   [2]

   (b) Explain why this polymer is suitable for the chair.   [6]

   (c) Use notes and diagrams to explain how the chair is
       manufactured.   [9]

   (d) Explain what health and safety measures the manufacturer
       would take to protect employees.   [5]

   (e) If the chair were to be made as a one-off prototype,
       name a material that you could use and explain why this
       would be suitable.   [6]

A2 exam question

1. Plastics are often used to replace traditional materials such as woods and
   metals in products today.

   Describe two different products that have been made with polymers
   to replace traditional materials. You should make reference to the
   materials used, the methods of manufacture and the benefits
   they bring to the manufacturer and the consumer.   [2 x 12]

Composite materials

Introduction to composites

Composites are produced by mixing together two (or more) different materials. The main advantage of doing this is that the properties from each of the materials can be enhanced and utilised.

For example, plastics have useful strength and rigidity with lightweight and good electrical insulation properties. These general characteristics can be enhanced by adding other materials. Adding plasticisers to polymers makes the polymers more processable by improving their flow properties (see p.5 for more details of additives for polymers). By adding strands (fibres) of glass to polyester resins, a very tough, rigid, lightweight material can be produced, i.e. glass-reinforced plastics (GRP).

Carbon fibre composites products are manufactured in a similar way to GRP. The carbon fibres – extracted from the polymer polyacrylonitrile by heating – are mixed with a resin, then heated in a mould to produce a composite that is much stronger than GRP. Its strength makes it suitable for producing the protective components for a modern Formula One racing car, or for high performance sports or aerospace products.

Kevlar is another special composite that has good protective properties due to the materials used and the way it has been processed (see pp.25–26).

Types of composite material

There are two main groups of composite materials:
- fibre-reinforced composites;
- particle-based composites.

The most important of these are the fibre-reinforced composites, since these are more commonly used in the manufacture of products.

The diagram below gives examples of materials found in both composite groups:

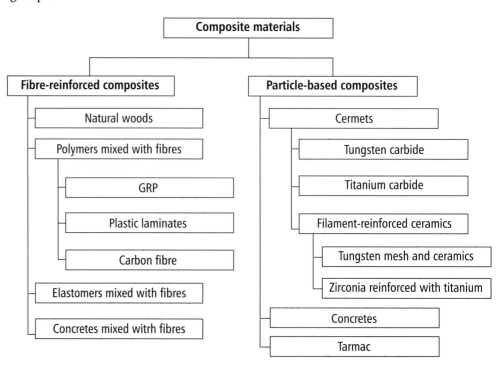

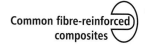

In addition to these materials, man-made boards can be considered to be composite materials. The term 'man-made boards' refers to those sheet or moulded materials where wood or wood fibre are bonded together to form a 'new' material.

Table 4: A range of man-made boards

Type of board		Common uses
Plywood		Backs of furniture, e.g. cabinets, bottoms of drawers, panelling; can be flexible for producing curved shapes
Block board		Generally used for tabletops and furniture carcasses
Stirling board		Flooring for sheds and workshops; also used for roofing and shuttering for casting concrete
Chipboard		Knockdown furniture, kitchen cupboards and worktops; usually veneered or laminated for furniture; also used for flooring
Medium Density Fibre board (MDF)		Furniture sides acting as a base for veneers; pattern making for castings
Hardboard		Backs of cupboards and drawer bottoms of kitchen units; can be supplied pre-coated

Advantages of man-made boards
- They have increased stability against warping.
- They have equal strength in all directions – unlike natural timbers.

Further reading
If you wish to study man-made boards in greater depth, a useful source would be:

- Focus CD-ROM *Resistant Materials 2*

Common fibre-reinforced composites

Here are the general characteristics of fibre-reinforced composites.
- They have a good strength/weight ratio (i.e. light in weight (low density) but strong compared to their weight).
- They are resistant to corrosion.
- They have a good fatigue resistance.
- They possess a low thermal expansion.

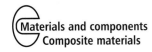

Glass fibre

- Glass used in glass-reinforced plastic (GRP) is spun to produce a fibre that is then coated to aid bonding to the resin.
- Fibres of glass are available in a variety of gauges (thicknesses) from coarse (30 μm) to very fine (5 μm). *(Note: 1 μm = 0.001 mm)*
- A mould is required for GRP. This can be produced quite cheaply from a range of materials including woods, metals and polymers.
- 'Laying up' is the term used for the processes involved in manufacturing with GRP.
- Inserts, e.g. mounting plates for securing fixtures, etc., can be included as the material is being 'laid up'.

Stages involved in the use of GRP

1. Coat the mould all over with a releasing agent.

2. Cut up the glass fibre matt into the minimum number of pieces that will cover the mould in three laminations. Add colour to the gel-coat and then the hardener to catalyse it.

3. Wearing polythene gloves, apply the gel-coat to the mould with an even brushing action to achieve a thickness of about 1 mm. Gel-coat is thixotropic and will not run.

4. When the gel-coat has cured, after about 30 minutes, coat it with a layer of catalysed lay-up polyester resin. On to this lay the first lamination of glass fibre matt. Stipple the matt using a stiff brush until it is thoroughly wetted and all air is driven out. Repeat with successive layers. Use surfacing tissue as the final lamination for an improved surface.

5. Leave for about 40 minutes while you clean all brushes and tools thoroughly. After this time, the edges can be carefully trimmed using a sharp knife.

6. Wait at least another 3 hours before separating the work from the mould. Full hardness is achieved after curing in approximately 24 hours, after which time it will be possible to work with wood- and metal-working tools.

Finished pond liner

Carbon fibre matting

Task

GRP matting is laid up with each layer lying across the one below, rather than all the fibres running in the same direction.

Explain what the likely outcome would be if the fibres did all face the same way.

Carbon fibre

This well-known material has come to the fore in recent years in its association with a variety of sports, e.g. Formula One racing cars, tennis racquets, fishing rods, etc.

The carbon fibres used in this material are produced by heating polyacrylonitrile filaments through a range of temperatures up to 2000 °C. This process ensures a high strength, lightweight material.

Initially carbon-fibre based products were expensive to produce due to the processes required to produce the fibres. However, as these processes continue to develop, manufacturing costs will continue to fall.

Processing with carbon fibres

Stages of the process:

Step 1   Carbon fibres are available in the form of woven matt, which is cut to the shape of a pattern using ceramic scissors.

Step 2   The material is placed into a mould half, where it is impregnated with resin and forced into the shape of the mould.

Step 3   The mould halves are fixed together and the whole lot is placed in an oven, where it is held at a temperature of 170 °C for up to 8 hours to promote the rigid cross-links in the resin.

F1 cars manufactured with carbon fibre reinforced polymers

Kevlar

Laying up carbon fibre in the mould

Loading the mould into an autoclave

Kevlar is a mixture of aromatic and aramid (nylon-like) molecules. These are melted and spun into fibres. The long-chain molecules are held together by strong hydrogen bonds. These fibres are always oriented parallel to the length of the fibres giving the material its very high strength. Weight for weight, Kevlar is five times stronger than steel and about half the density of fibreglass.

How does Kevlar work?

Kevlar fibres are woven into a 'cloth', which can then be fashioned into protective equipment. It protects the wearer by operating essentially as a net – in the same way that a goal net will absorb the force of a football being kicked into it. All the horizontal and vertical fibres absorb some of the impact. Kevlar can absorb impacts that would cause serious injury to an unprotected human body. The wearer is still affected by the impact, but saved from serious injury.

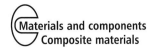

Arrangement of Kevlar fibres

Uses of Kevlar

Kevlar is used in a wide variety of applications because of its unique properties, including:

- body protection, such as bullet-proof vests where lightweight properties, comfort and flexibility are important;
- sports equipment, such as skis, helmets and racquets, where lightweight properties and strength are important;
- sails for windsurfing, where the material has to withstand high speeds;
- run-flat tyres that will not damage the wheel rim;
- gloves for use in the glass and sheet-metal industries.

Kevlar body armour

General properties of Kevlar:

- high strength to weight ratio;
- low electrical conductivity;
- high chemical resistance;
- high toughness;
- high cut-resistance;
- flame resistant and self-extinguishing.

Task

Using the Internet, research into Kevlar and complete the following:

(a) Write a brief summary of what Kevlar is made from.
(b) Describe two different products that Kevlar is used in and explain why it is used for these.

Task

Kevlar is used in the manufacture of arm protectors for use by people in the glass industry.

State the reasons why Kevlar is an appropriate material for this application.

Applications for fibre-reinforced composites

Table 5: Typical uses of fibre-reinforced composites

Composite	Common uses
Glass-reinforced plastic (GRP)	A mixture of glass fibres and polyester resins, used in: the manufacture of some vehicle bodies; the front sections of some locomotive engines and sports equipment, e.g. canoes and boat hulls
Carbon fibre	A mixture of carbon fibres and resin, used in: the manufacture of sports equipment, e.g. tennis racquets, bicycle frames, etc.; the manufacture of artificial limbs
Kevlar	A mixture of aramid and aromatic fibres interwoven into a cloth, used in body armour and sports equipment
Plastic laminates (Tufnols)	Cotton/resin composites, used in the manufacture of gears and cams giving quieter operation than metal components and greater wear-resistance
Plastic laminates	Mixtures of resins and paper, used to produce worktops for kitchens

Further reading

Useful resources on fibre-reinforced composites include:

- *Materials for the Engineering Technician*, *Third Edition,* R A Higgins (Arnold)

- **www. mdacomposites.org** covering manufacturing with FRP (fibre-reinforced plastics) and thermosetting materials

Reinforced concrete

Concrete is classed as a particle-based composite. Concretes are generally very good when subjected to compressive loads as, for example, in foundations for a building, but very poor when in tension, for example, when used as a beam when spanning a distance.

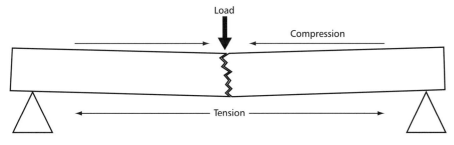

Non-reinforced concrete may crack under tension

We are looking at this material in the fibre-reinforced composites section because, in order to combat the possibility of failure under tension, reinforcing bars can be placed in the concrete shuttering prior to casting the concrete. These reinforcing bars will then become surrounded by, and gripped by, the concrete.

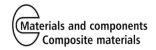

Casting concrete over reinforcing bars

The inclusion of reinforcing bars has led to the design of longer spans in bridges and buildings. However, there is still some potential for cracking in places where the beam/structure is under tension, i.e. on the underside of the beam.

This has led to a process where the reinforcing bars are put under tension prior to and during the casting of the concrete structure.

Once set, the tension is released on the bars having the effect of placing that part of the structure under compression. This means that the beam/structure is better able to withstand heavier loads or bridge longer spans. The diagram below shows the effect of the use of high-tensile reinforcing bars in concrete.

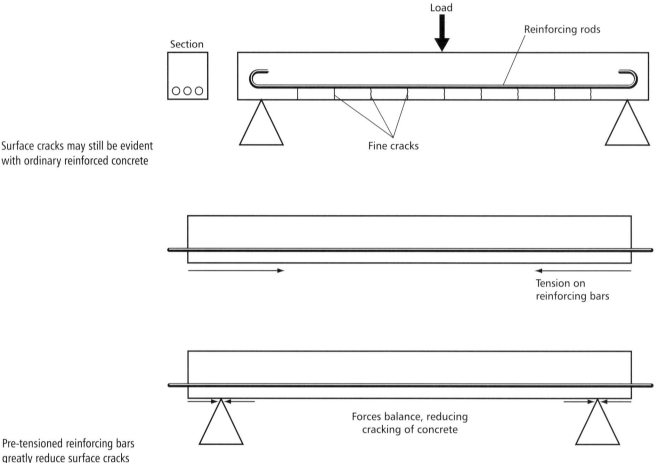

Surface cracks may still be evident with ordinary reinforced concrete

Pre-tensioned reinforcing bars greatly reduce surface cracks

Particle-based composite materials

The general characteristics of particle-based composites are:

- They have a high strength in compression, as in the case of concretes; less so in tension.
- They have good stability.
- They have a uniform structure ensuring consistent strength.
- They are generally free from surface defects.

Concrete

We completed the last section with a reference to reinforced concrete. Concrete (without the reinforcing) is appropriate for this section too.

Concretes are made up of materials known as aggregates, as well as sand and cement. The characteristics of the final concrete material are determined by the ratio of the constituent parts.

Task
Carry out some research into concretes and find the mixture ratio for the construction of a garden path or driveway.

Concrete is thoroughly mixed while it is dry. Water is then added. Mixing continues until every particle of aggregate and sand is coated in cement paste (this is the bonding agent). Once cast, the concrete is left to harden. During hardening, the temperature of the mix will rise – this is due to the chemical reactions that take place.

Advantages and disadvantages of concrete

Advantages
- It can be moulded into complex shapes.
- It has properties similar to stone.
- Components are more readily extracted than stone.
- It can be cast *in situ* (on site), whereas stone has to be quarried and cut to shape.
- It is good in compression.

Disadvantages
- It is poor in tension, making it necessary to reinforce the concrete when spanning large distances.

Cermets

Cermets are another group of particle-based composites. These are mixtures of both metal and ceramic particles. An example of a common cermet is tungsten carbide – a combination of the ceramic tungsten carbide and the metal cobalt. This material is used extensively for cutting tools, as it keeps its edge very well.

Other examples of cermets include a mixture of aluminium oxide and cobalt – used in the manufacture of components for jet engines.

Advantages and disadvantages of cermets

Advantages
- It is resistant to high temperatures.
- It is tough and shock-resistant.

Disadvantages
The very nature of the component materials of a cermet such as tungsten carbide (i.e. brittle, high melting point metal and ceramic materials) exclude processing in the same way as materials which melt at lower temperatures. Sintering, therefore, is one of the few processes suitable for cermets.

Further reading
If you wish to learn more about particle-based composites then a useful resource would be:

- *Materials for the Engineering Technician, Third Edition,* R A Higgins (Arnold)

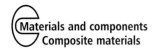

Products where traditional materials have been replaced by composites

Kitchen knives and flooring are two examples where traditional materials have been replaced by composites.

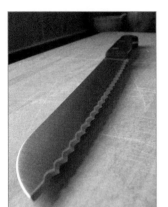

Tungsten-edged kitchen knife

Kitchen knives

Traditionally knives for use in the kitchen have been made from a range of materials, including stainless steel. They are very tough and corrosion-resistant but do not keep their sharp edge very well, resulting in the need for re-sharpening.

Modern processing techniques have enabled the fusion of tungsten carbide onto the cutting edge.

Advantages and disadvantages of tungsten carbide on the cutting edge

Advantages
- It results in a much more durable cutting edge.
- There is a reduced necessity for re-sharpening.

Disadvantages
- When the food is being cut, care must be taken not to cut further into materials below, e.g. the chopping board. The blade is hard enough to cut through ceramic glaze, most metals and certainly woods.

Traditional floorboards

Flooring

Flooring used in the construction of modern houses is now largely chipboarding, whereas traditionally it would have been solid wooden floorboards. The edges of the chipboard sheets are also tongued and grooved in the same way that floorboards are.

Advantages and disadvantages of chipboard flooring

Advantages
- Large areas can be covered with one sheet.
- The cost is reduced.

Disadvantages
- Chipboard flooring would not be left on display as floorboards could, but must be covered in a flooring product such as a carpet.

Tasks

(a) For each of the following composites, list as many applications that you can think of where the composites are used:
- glass-reinforced plastic;
- concrete;
- carbon fibre;
- Kevlar;
- laminated glass.

(b) Draw up a revision table for each of the composites from part (a). Your table should include details of the constituent parts (the ingredients) of each composite and the properties of the composite.

Carbon fibres in bicycles

This is a useful case study for A2 students who should be familiar with several specific product applications for composites.

One of the most famous applications of carbon-fibre reinforced plastic is in the development and manufacture of the Lotus bicycle, designed by Mike Burrows and ridden by Chris Boardman in the 1992 Barcelona Olympics. Boardman achieved the gold medal.

Burrows, an engineer, cyclist and racer of recumbent cycling machines, had been using composites such as Kevlar, GRP and carbon fibre to make the aerodynamic body shells for his recumbent bikes. Realising the immense strength, low weight and ease of forming these materials, Burrows set about developing a new type of frame for a racing bike.

The traditional aluminium alloy bike frame is quite heavy when compared to composites such as carbon fibre. The aluminium alloy frame is made in a tubular construction and the tubes, being relatively wide, offer resistance to airflow (called drag). This slows the bike down.

Burrows produced a prototype monocoque (single-piece shell frame) by laminating carbon fibre over wooden moulds, producing two halves, which were then glued together with resin and reinforced with more carbon fibre. The resulting frame was very lightweight and much thinner than a traditional tubular bike frame, whilst still maintaining the strength of an alloy frame. This thin, almost wing-like frame, offered considerably less wind resistance. In making the prototype, Burrows used the expertise of Lotus – famous for performance sports cars – in their use of advanced materials such as carbon fibre. After Burrows added the mechanical parts to the frame, Lotus conducted wind tunnel experiments with Boardman to find the riding position that would offer least wind resistance. Carbon fibre was also used to manufacture a rather striking, cone-shaped helmet that Boardman used to help the air flow over his head and down the line of his back.

Chris Boardman aboard the Lotus Superbike

The resulting design helped cut seconds off Boardman's lap time, and he comfortably took gold at the Olympics. This revolutionary bike design is now used by other bicycle racers world-wide.

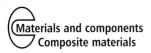

PRODUCT ANALYSIS EXERCISE: *glass-reinforced plastics*

1. Glass-reinforced plastic is often used in products such as boats, sports equipment, water tanks and car body panels. Explain why GRP is suitable for such products.

A product made from composite materials

2. Use notes and diagrams to explain how such products are manufactured using GRP.

3. An alternative to GRP might be carbon-fibre reinforced plastic. Explain the advantages of using this.

4. Materials such as FRP and GRP are known as composites. Define the term 'composite'.

5. Describe the health and safety precautions you would take when using composites such as FRP and GRP.

Exam questions

AS exam question

1. (a) For each of the following composites, describe why they are suitable for the applications listed.

(i)	MDF	Flat-pack furniture
(ii)	Concrete	Paving slabs
(iii)	Carbon-fibre reinforced plastic	Fishing rods

[3 ×6]

(b) Use notes and diagrams to show how a product, such as a boat or vehicle body panel, is manufactured using glass-reinforced plastic (GRP). [10]

A2 exam question

1. Combining two or more materials can produce a composite with improved properties compared to the original materials.

(a) Referring to a specific product with which you are familiar, describe how composites are used in its manufacture and explain the benefits of the use of composites in the product. [12]

(b) For a specific composite used in school workshops, carry out a risk assessment to identify the hazards of its use and what control measures you should take. Your risk assessment should indicate who is at risk and the level of risk. [12]

Metals

Introduction to metals

Along with woods, metals have been in use for a thousand years or more and are seen as traditional materials. The passage of time has seen these materials develop into a wide range of metals and alloys, all with a variety of useful properties and characteristics.

At AS level
As an AS level student, you need to learn and understand the range of metals available, their general characteristics and properties along with examples of uses.

You also need a good level of understanding of how metals can be processed – this will also include the use of some basic heat treatments.

Both AS and A2 level students should be aware of corrosion processes that can affect metals.

At A2 level
A2 students should build on their AS level work to understand why metals have been used to manufacture a particular product, and also why metals have been replaced in some products by different materials.

The pages on sources of metals and their manufacture are intended only as an overview of how these materials are obtained and processed into a usable form. You should refer to your exam specification to find out how much you need to learn about these processes.

Types of metal

The range of metals available can be classed as either ferrous or non-ferrous. Within these two groups, metals can be further separated into alloys or non-alloys. The diagram below shows how the more common metals are related.

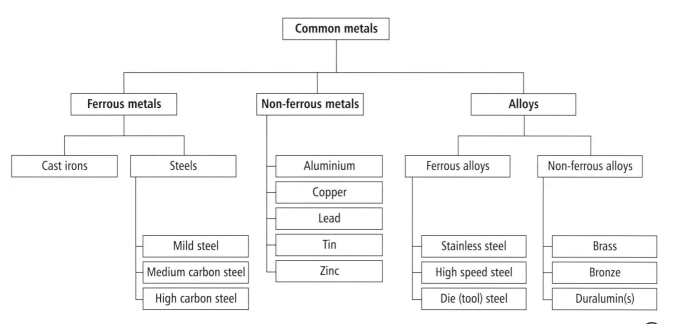

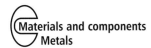
Sources of metals

Gold is the only metal that is found in a usable form; all other metals are found as ores. Table 6 shows the ores for individual metals.

Table 6: Common metals and their ores

Metal	Ore
Iron	Magnetite, haematite
Copper	Chalcopyrite
Aluminium	Bauxite
Lead	Galena
Tin	Cassiterite
Zinc	Zinc blende

Availability of metal ores

- 25% of the Earth's crust is made up of metal ores.
- Aluminium is the most common ore, followed by iron.
- The more rare the material the more expensive it is, generally.

However, some of the more common ores, e.g. aluminium, can be expensive to process.

Microstructures of steel, brass and aluminium

Normal structure of a mild steel (the dark areas represent carbon)

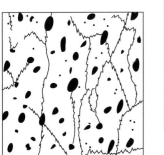

Structure of brass showing black specks of zinc

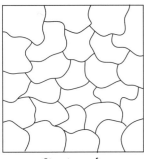

Structure of aluminium (pure)

The structure of metals

All metals are made up of crystals. Each crystal has a boundary that is firmly bonded to the boundary of a neighbouring crystal.

The nature of the crystal depends very much on the material. For example, steel is made up of iron and carbon so these elements will be seen within the microstructure of the material.

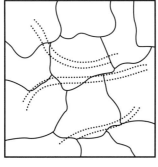

Crystals that have not been stressed by working or reformed by annealing. The continuous lines represent crystal boundaries; the dashed lines repesent dislocations (faults) in the structure

Further reading

If you would like to learn more about the sources and structure of metals then a useful resource would be:

- *Engineering Metallurgy Part 1 – Applied Physical Metallurgy, Fifth Edition*, Higgins (Hodder and Stoughton)

Ferrous metals: iron and steel

Iron is produced directly from its ore through the use of a blast furnace. The material that is produced is called 'pig iron' and is not of a sufficiently high quality to be of any commercial use.

Pig iron is 'converted' into steel by the introduction of carbon into its structure. This process is carried out in a basic oxygen furnace.

An alternative steel-making process is the electric arc furnace – more often used in the production of specialist steels.

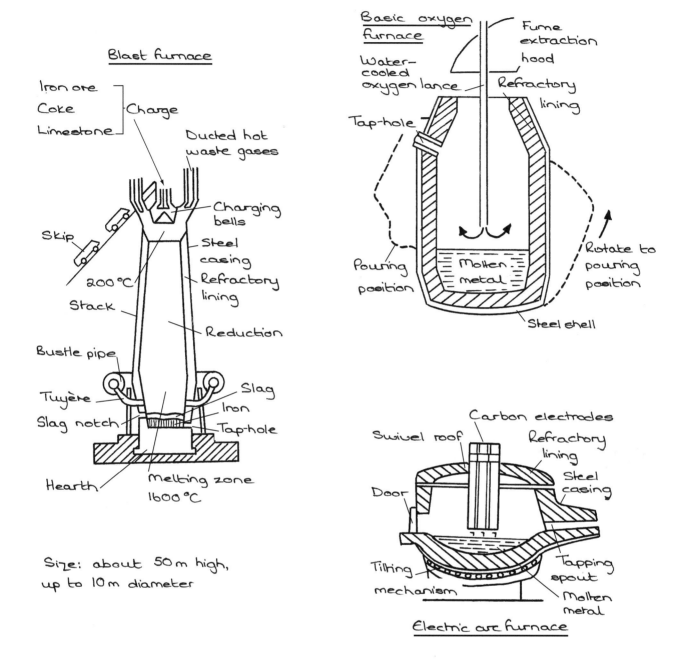

Iron and carbon

Iron is generally soft and ductile, which does not make it a very commercially useful material. When carbon (a very hard, brittle element) is mixed with the iron the characteristics (properties) of iron are greatly improved, the result being a harder and tougher material – steel (see Table 7 overleaf). Increasing the carbon content has the following effects.

- The material becomes harder (i.e. the effect increasing in the direction of the arrow).
- Toughness reduces and, indeed, cast iron can be brittle under impact.
- Both medium and high carbon steel can be heat-treated to make them even stronger and harder, so producing materials that are of sufficient hardness and strength to be formed into cutting tools. Mild steel, while containing some carbon, has insufficient carbon in its structure to enable it to be heat-treated in the same way.

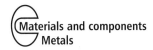
Table 7: Types of steel

Base material	Additional element, carbon	Type of steel	Ductility	Hardness	Toughness
Iron	<0.3%	Low carbon steel (mild steel)			
	0.3–0.6%	Medium carbon steel			
	0.6–1.7%	High carbon steel			
	3.5%	Cast iron			Brittle

Direction of arrows indicates increase in property.

Applications for ferrous metals

Table 8: Typical uses for steels and cast irons

Name	Common uses
Mild steel	Nuts, bolts, washers, car bodies, panels for cookers and other white goods
Medium carbon steel	Springs, general gardening tools
High carbon steel	Hand tools, scribers, dot punches, chisels, plane blades
Cast iron	Machine parts, brake discs, engines

Non-ferrous metals

Non-ferrous metals do not contain iron. This group of materials includes aluminium, copper, lead, zinc and tin; as well as precious metals such as silver, gold and platinum.

Although aluminium ore (bauxite) is the most abundant ore in the Earth's crust, aluminium is not the most processed metal – steel is. This is because aluminium is more difficult to process, consuming large amounts of energy. It is therefore more costly to produce aluminium.

The production of copper from chalcopyrite is a similarly expensive process requiring the ore to be crushed, followed by a number of refining processes to remove other metals from the ore.

Both aluminium and copper require processing by either an electrolytic process or by re-melting; these processes are also required for materials such as tin and zinc.

Table 9: Non-ferrous metals

Metal	Melting temperature	Common uses and properties
Aluminium	660 °C	Kitchenware, such as saucepans; when drawn into wire, used in overhead power cables – it is an excellent conductor of electricity
Copper	1083 °C	Electrical contacts, domestic pipe work for central heating and water; in wire form, it is used for electrical cable and wire; also used in jewellery
Gold	1063 °C for fine gold	Primarily thought of as a metal for jewellery, but also has applications in electronics in the form of contacts for switches and credit/telephone SIM cards
Lead	330 °C	A very soft but heavy metal used for flashing between roofs and adjoining brickwork; very durable

Metal	Melting temperature	Common uses and properties
Platinum	1755 °C	Used as a precious metal in the manufacture of jewellery; is also used in wire form to produce thermocouple cables
Silver	960 °C for fine silver	Used for many years in the manufacture of expensive cutlery and various decorative items; also used in the processing of photographic film
Tin	232 °C	Rarely used in its pure state, but applications include food wrapping (foil) and coating for steel plate in the manufacture of food cans
Titanium	1675 °C	Has a good strength/weight ratio and is a very clean material, making it suitable for surgical applications such as hip replacements; also used for spectacle frames
Zinc	419 °C	Used as a coating for steels, i.e. galvanised steels; used for the manufacture of products such as buckets, and casings for electrical units; can be die-cast to produce high detail products, such as lock mechanisms and small gears

Further reading

For those interested in jewellery and the metals used in the manufacture of jewellery, a useful resource is:

● *Jewellery Making Manual*, Sylvia Wickes (Little, Brown and Company)

Alloys and alloying

In the same way that composite materials (p.22) take the best from both materials for the proposed application, the alloying of metals achieves a similar result, producing materials with enhanced properties.

Individual metals have a limited range of properties that can only be enhanced by heat-treating them in some way. To obtain a better range of properties and characteristics two (or more) metals can be mixed together to produce an alloy.

For example, the addition of zinc to copper produces a much harder and stronger material than pure copper. Alloying changes other characteristics of the material. Mixing copper with zinc to make brass, for example, changes the colour of the metal to a yellow/gold making the material attractive to purchasers.

Table 10: Common alloys

Name	Base metal	Composition	Common uses
Duralumin	Aluminium	4% copper 1% manganese 0.1% magnesium	Structural components for aircraft
Brass	Copper	35% zinc	Cast valves and taps, boat fittings and ornaments
Bronze	Copper	10% tin	Statues, coins, bearings

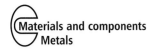

Benefits of alloying

In general, the benefits of alloying metals are:

- changes the melting point;
- changed colour;
- increased strength, hardness and ductility;
- enhanced resistance to corrosion and oxidation;
- changed electrical/thermal properties;
- improved flow properties producing better castings.

> **Task**
>
> List the benefits of using brass for the 'pins' on a three-pin domestic plug. Why would copper be less suitable?

Alloying steels

High Speed Steel (HSS) in action

Alloying steels with elements like chromium and nickel will produce stainless steel – a well-known group of metals with good corrosion resistance, good hardness, strength and toughness.

Most metals, steel included, will become less hard and more ductile when heated. By alloying with tungsten, chromium and cobalt, a range of 'high speed steels' can be produced, which do not lose their cutting edges when working at high temperatures (Table 11).

Table 11: The effects of alloying steels with other elements

Alloy steel	Alloyed with	Characteristics	Common uses
Stainless steel	Chromium, nickel, magnesium	Tough and wear-resistant; corrosion-resistant	Sinks, cutlery, sanitary-ware, equine accessories
High speed steel (HSS)	Tungsten, chromium, vanadium	Very hard, will cut while at red heat	Cutting tools, such as drills
Tool and die-steels	Chromium, manganese	Very hard and tough, with excellent wear-resistance	Fine press tools, extruder dies, blanking punches and dies, some hand tools
High tensile steels	Nickel	Good tensile strength and toughness, generally corrosion-resistant	Car engine components

Further reading

A useful resource of further information about alloying of metals is:
- *Materials for the Engineering Technician, Third Edition,* R A Higgins (Arnold)

Work hardening and heat-treating metals

Work hardening is a phenomenon peculiar to metals. Work hardening occurs when the material is 'cold-worked' by, for example, bending, rolling, hammering and drawing.

The processes involved when the material is being worked results in the distortion of the crystals to the point where they become highly stressed, making the material 'harder' in that area.

> **Task**
>
> Straighten a paper clip. Now create a sharp bend in the metal. Now attempt to bend the material back. What happens?

Heat treatments

Heat treatments are those processes of heating and cooling metals in a controlled way, in order to achieve a beneficial change in the properties of the material. The more common heat treatments are:

- annealing;
- hardening;
- tempering;
- normalising.

Annealing

Annealing is a heat treatment that reverses the internal stresses associated with work hardening. It is achieved by heating the material to a temperature where the crystals grow, making the material softer and more ductile. The temperature must be maintained for a sufficient amount of time for the material to 'soak' at that temperature. The material should then be allowed to cool very slowly.

Hardening

This heat treatment changes the way the carbon within the steel affects the strength and hardness of the material. When medium carbon steel, for example, is heated to a specific temperature, the carbon in the structure moves out of its normal position.

If the material is then 'quenched' the carbon does not have sufficient time to move back to its original position and so causes internal stresses, which serve to harden, and strengthen, the material.

Tempering

Tempering is the heat treatment that follows the hardening of medium and high carbon steels. If left without being tempered the product that has been hardened will, potentially, be quite brittle – so that with sufficient mechanical shock the product could well fail by cracking or shattering.

Tempering then reduces the amount of brittleness caused by hardening. The materials' internal stresses are reduced, by allowing the atoms and molecules that make up the structure of the material to 'relax' a little.

The amount of tempering given to the material depends on the use; Table 12 (overleaf) gives some examples.

As the tempering temperature rises:

- the material's hardness is reduced;
- toughness is increased.

Distorted crystals

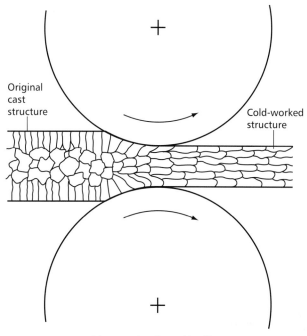

Original cast structure

Cold-worked structure

Diagram showing cold-rolling

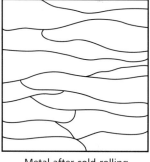

Metal after cold-rolling

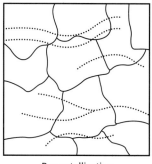

Recrystallisation

---

* Quenching

Quenching is the term given to the rapid cooling of heat-treated components. Different media can be used for quenching.

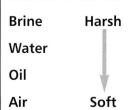

Brine   **Harsh**

Water

Oil

Air   **Soft**

The amount of agitation given to the product in the media also determines the rate of cooling and prevents rapid local cooling, which may result in cracks.

Table 12: Tempering temperatures

Colour			Temperature	Uses
Pale straw	harder	less tough	230 °C	Lathe tools
Straw			240 °C	Drills and milling cutters
Dark straw			250 °C	Taps, dies, punches, reamers
Brown			260 °C	Plane irons, lathe centres
Brown-purple			270 °C	Scissors, press tools, knives
Purple			280 °C	Cold chisels, axes, saws
Dark purple			290 °C	Screwdrivers
Blue	hard	tougher	300 °C	Springs, spanners, needles

Drills are brittle and snap easily

Normalising

Normalising is a heat-treatment process where the crystal structure is allowed to become uniform, i.e. the crystals become a similar size.

The process includes heating the piece of work (e.g. a forged component) to a specific temperature where the crystal structure begins to change. This temperature is maintained – 'soaked' – until all of the crystals have been refined. The workpiece is then cooled in air, to produce a material with additional toughness and greater ductility.

Age hardening

Age hardening is not a process in the conventional sense of the word, since no heat is applied to the metal nor is it manipulated in any way. It has been included in this section simply because it is utilised to enhance the characteristics of the material.

The phenomenon of age hardening applies generally to aluminium/copper alloys (duralumins). After heat-treating the material is left for a period of time during which the internal structure moves slightly, making the material stronger and harder.

Task

Describe the general characteristics of metals and their alloys. For example, what benefits does stainless steel have over mild steel, or what are the benefits of bronze over copper?

Case hardening

Case hardening is a method of increasing the hardness of steels that do not have sufficient carbon content to affect internal hardening. This basically refers to mild steels (i.e. those with less than 0.3% carbon) and involves the addition of carbon to the outer skin of the material. This outer surface can then be hardened leaving a tough core. Examples would include cams, where the surface needs to be resistant to the wear being imposed on it while the rest of the material needs to be tough and shock-resistant.

There are a number of techniques for achieving this, and all of them involve the component being placed into a carbon-rich atmosphere while being heated to around 950 °C. At this temperature carbon atoms are able to enter the material's structure, building up the carbon content at the surface of the material. The longer the component is left in the carbon atmosphere, the thicker the carbon layer produced.

Carburising is one such method.

Carburising

The component to be case hardened is placed in a ceramic box packed with carbon-rich material, and then heated for a predetermined length of time to produce the thickness of carbon layer required.

Following case hardening the product must be heat treated to ensure the surface is hardened. Here, the material is heated to around 760 °C, then quenched to produce the hard case.

Advantages and disadvantages of case hardening

Advantages

- Steels that do not have sufficient carbon for heat-treating in the same way as steels with higher carbon contents can be given a hardened surface.

- This process leaves a tough inner core, making the material suitable for products such as gears, steering components and camshafts for cars: all of which take a lot of wear.

Disadvantages

- In the majority of cases, grain-growth occurs. This needs a machining process, such as surface grinding, to return the material to its required size.

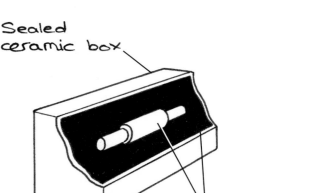

Sealed ceramic box

Mild steel product encased in carbon-rich material

Case hardening

Nitriding

Nitriding is another case-hardening technique that involves immersing the product in the hardening medium for a specified time while being heated, this time to about 500 °C. The medium in this case is nitrogen, while the materials that are case hardened in this way are special steels containing aluminium, chromium and vanadium. Products hardened in this way include components for aero-engines.

Advantages and disadvantages of nitriding

Advantages

- No additional hardening is necessary.
- It removes the chance of cracking on the surface.
- It increases resistance to corrosion.
- It's a clean process.
- It's economical for large numbers.

Disadvantages

- Initial set-up costs are high.
- If over-heated there is a permanent loss of hardness.

Other hardening methods

Other methods of hardening the surface of a component include flame hardening and induction hardening. Both of these techniques rely on the material used having a carbon content of 0.4% or more, and involve heating the surface only followed by rapid cooling by a jet of water. Both of these processes can be mechanised and are used for the surface hardness of gears and camshafts.

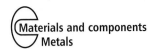

Advantages of flame and induction hardening

- Localised areas of the product can be hardened, leaving those areas unaffected where toughness is required.
- Grain growth does not occur with this method of hardening; so additional machining is not required following hardening.

Further reading

A useful resource of information about traditional heat-treating methods is:

- *Hardening, Tempering and Heat Treatment,* Tubal Cain (Argus Books)

Processing metals

There are numerous processes available for manufacturing products from metals. These processes can be divided into three main areas:

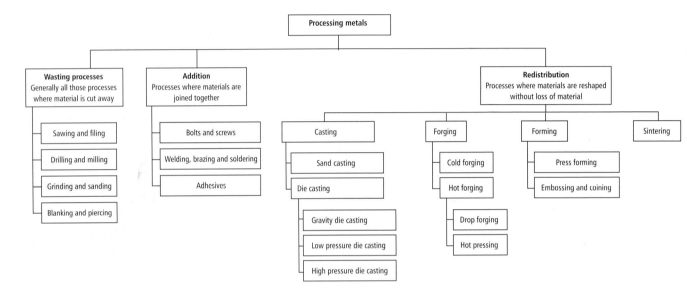

It is clear that a number of processes will relate equally well to a number of different types of material. For example, wasting processes like sawing and filing apply equally to woods as they do to plastics as well as to metals. In this section, we will only look at those processes that relate to metals.

Wasting processes (relating to metals)

Blanking and piercing

Sheet metals can be cut to a required shape using punches. These cut through the material using a shearing action – much in the same way that scissors cut through paper.

A guillotine is usually used to cut sheet metal off a roll into usable sheet sizes. These sheets are then passed into either manually operated or automatic machines that will cut the material to shape and/or punch holes into it.

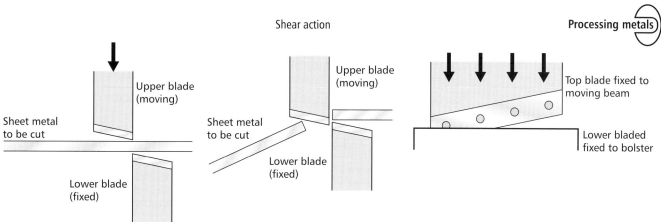

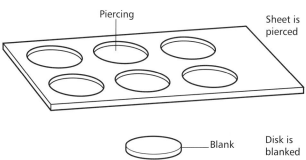

Piercing

Sheet is pierced

Blank

Disk is blanked

Diagram showing blanking and piercing

When a sheet of metal has a hole punched into it, it has been pierced. When the piece that has been punched out of the sheet is to be used, it is called a 'blank'.

Products, such as soft drinks cans, are made by punching disc-shaped blanks from the sheet material. The process is set up to maximise efficiency with as little waste metal left as possible.

Some products require both blanking and piercing, e.g. casings for desktop computers.

Other cutting processes

Heavyweight products such as armoured tanks, earthmovers and diggers, have components cut from steel sheet that is between 8 mm and 12 mm thick. The normal action of a guillotine or hydraulic punch is inadequate in this case, so other cutting processes are used.

Profile cutting using oxy-acetylene torches has been used for a number of years in, for example, the shipbuilding industry, for cutting large thick sheets of steel to shape prior to welding into position in the construction of a ship.

More modern techniques include plasma cutting and the use of lasers to produce sufficient energy to cut through the thick sections required for diggers and earthmovers, for example.

Plasma cutting

Plasma cutting uses an electric arc to generate the heat energy required, plus the energy of either compressed air or an inert gas such as argon to blast through the material. This process produces very little waste material. A fine cut is achieved with little or no refinishing to remove burrs.

Laser cutting

Laser cutting can produce profiles of much finer detail. The width of cut is much narrower than that of plasma cutting, resulting in even less waste material. Laser cutting, as well as plasma cutting, can be automated using fully controlled CNC machines resulting in components of consistent quality.

The results of plasma cutting (top) and laser cutting (bottom)

Processing by redistribution

Forming materials to shape can be achieved by a number of processes, the use of which depends on the type of product being produced and the nature of the material being used.

Processes associated with redistribution techniques include:

- press-forming;
- forging;
- casting;
- moulding.

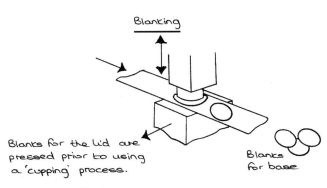

Blanks for the lid are pressed prior to using a 'cupping' process.

Sketch showing blanking prior to press-forming

These processes include hot- and cold-working techniques. The term 'redistribution' refers to the shaping of the material either cold, heated or as molten material. Examples include the sheet metal forming of a drinks can or car body, the forging of heated metals into shape to produce coins, tools or axles for vehicles, and the casting of molten metals to produce such varied products as sculptures or disc brakes for cars.

Press-forming

Press-forming is carried out with the material at room temperature. The process relies heavily on the ductility of the material being pressed. If insufficiently ductile, the material may have to be annealed to increase its ductility.

Sketch showing sheet metals being pressed

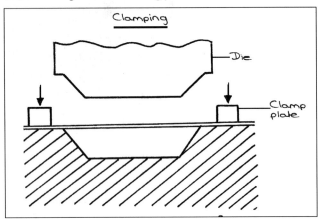

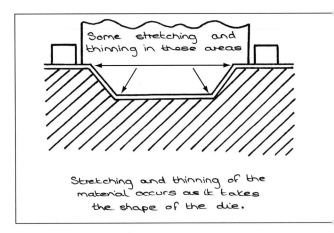

Stretching and thinning of the material occurs as it takes the shape of the die.

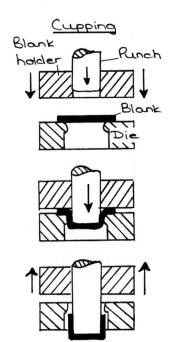

Press-forming is carried out using a punch and a die which are both manufactured from toughened die-steel; this makes them resistant to impacting loads and wear from contacting the material being pressed.

Car body panels are pressed from mild steel sheet to produce the vehicle's overall shape once assembled. The complex shapes produced require the generation of very high stresses to overcome the resistance of the material being pressed.

There are advantages of pressing a sheet material to a more three-dimensional shape, including that of greatly increased stiffness. This in effect has the benefit of reducing the amount of material necessary to build the vehicle to a good safety standard.

In addition to forming to shape, press tools can also incorporate shears to cut sections away. If we look again at the completed car body panel, we can see the holes have been cut to form door pillars and windows.

Other examples of press-formed sheet materials include domestic radiator panels, kitchen products such as meat trays, and cooker tops.

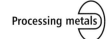

Domestic radiator

Press-formed car body panel

Embossing

This process is similar to press-forming in that it is used to change the three-dimensional shape of the sheet material. The main difference is that embossing is used to create decorative features with some quite intricate detail.

Examples of embossing include jewellery, confectionary tins, paper and card products such as letter-headed paper and greetings cards.

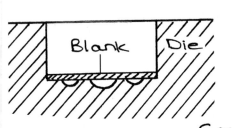

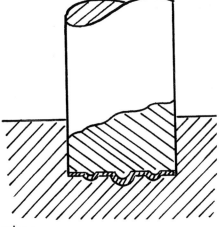

Embossing

Example of embossing

Deep drawing

Forming sheet metals into the required shape means the material undergoes a number of processes. Exactly which processes used will be determined by the product being made.

Continuing with the example of the soft drinks can (see p.43), the blanked discs are formed into a cup shape using, as the name suggests, the 'cupping process'.

In order to obtain the elongated shape of the can, the cupped shape is then deep-drawn. This requires the material to be pushed through a series of forming rings, which employs the material's property of ductility allowing the material to be drawn out without fracture.

- A phenomenon of deep drawing is reduction in the wall thickness of the can. It is quite possible that the wall thickness has been reduced to a third of the thickness of the base – to accommodate the elongation of the side.
- During deep drawing, the base of the can is formed by press-forming.
- Other processes are used to shape the neck of the can prior to filling and sealing the top to the can body.

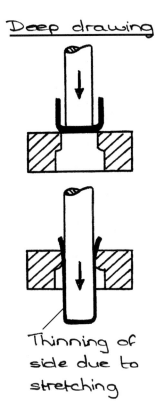

Deep drawing

Thinning of side due to stretching

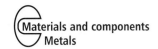
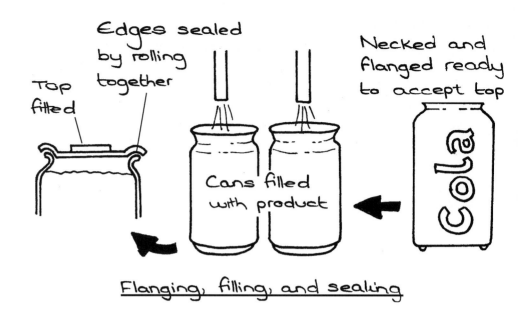

Top fitted

Edges sealed by rolling together

Necked and flanged ready to accept top

Cans filled with product

Cola

Flanging, filling, and sealing

Task

Create a series of sketches showing how a stainless steel kitchen sink may be produced. Use the stages shown above as a guide.

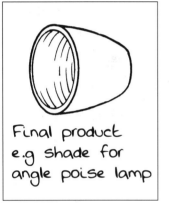

Final product e.g shade for angle poise lamp

Sketch showing spinning

Benefits of forming

Car bodies are good examples of products manufactured by forming, in this case using press-forming.

Spinning is a traditional process used for forming three-dimensional hollow objects from flat sheet metals. The process involves a measured circular sheet of material – this can be aluminium, brass, copper, mild steel or stainless steel – which is gripped in a machine similar to a lathe.

One of the supports for the material is the former around which the material is to be forced to 'flow' as the diagram left shows.

Products manufactured in this way include saucepans and woks, prior to having a handle fitted. The tell-tale signs of spinning are the concentric lines around the outside of the product – denoting the path of the tool.

Advantages of press-forming

- Sheet metal – for most car bodies this is mild steel formed to shape using dies. This gives the material a more three-dimensional shape, either through simple punching and folding or by stretching into a curved shape.
- The act of folding a material gives that material greater stiffness and rigidity (try the task below). A similar phenomenon is seen when a sheet of mild steel is forced into a 'cup' shape, which again is capable of supporting its own shape as well as additional forces.
- The material has been stretched – requiring good ductility in the material. By being stretched, the worked material hardens (see section on heat treating metals, p.39); since hardness is a by-product of strength, the material's structural strength increases.

So a car body achieves its strength and rigidity by the joining together of a large number of press-formed mild steel components to form a monocoque shape, which does not require an additional chassis. At the same time, weight is kept to a minimum.

Task

Try this for yourself. Take a sheet of paper. On its own it is quite flimsy; now either roll it into a tube or crease it into a fold (or folds). What has happened?

Casting processes

 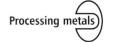

Like polymers, metals can be heated to known melting temperatures. When molten, the liquid metal can then be poured (or forced under pressure) into a mould. Moulds can be created from sand, alloy steel or ceramics, depending on the metals being cast.

Sand casting

In this process, sand is used for the moulds. The sand is specially prepared, to contain oils that act as binders to help it hold its shape while the hot metal is being cast into it.

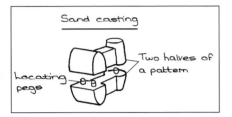

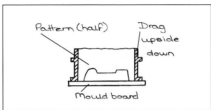

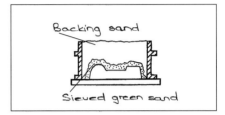

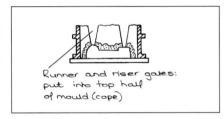

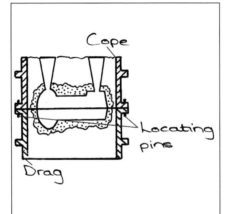

Stages of sandcasting process

Stages of the process:

Step 1 A 'pattern' is made – this could be made from a range of materials such as: woods (like yellow pine, jelutong); metals (such as aluminium); polymers (like polystyrene). Patterns can be split for more complex shapes.

Step 2 Each half of the pattern is placed on a baseboard. A mould box half is placed over it.

Step 3 Green sand is 'tamped' around the pattern forcing it into contact with the pattern. This is followed by backing sand (usually recycled sand).

Step 4 The pattern is removed from the mould half. The runner and riser gates are then cut into the top half of the sand mould.

Step 5 The mould halves are fitted together with locating pins – ensuring correct alignment.

Step 6 The molten metal is poured into the running gate. The riser is used to indicate when the mould is full. De-gassing tablets may be necessary to reduce the risk of a porous casting.

Step 7 Once the metal has solidified, the sand mould is broken open leaving the product with runner and riser gates attached. These will be removed either by using a band saw or by some other means, depending on the material being cast.

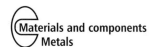

Task

Give at least three examples of sand-cast products.

State the material used to manufacture each of the products. Identify further manufacturing processes that the products have undergone.

Advantages and disadvantages of sand casting

Advantages
- Complex three-dimensional shapes can be produced.
- Cores can be used to produce hollow sections.
- It's appropriate for small runs.
- Automated processes are suitable for longer production runs.

Disadvantages
- Due to the poor surface finish, some machining will be necessary.
- It's not as accurate as die or investment casting.
- It has a low rate of output and is therefore suitable only for small production runs.

Table 13: Metals used in die casting

Metal used	Melting temperature
Aluminium	660°C
Magnesium	650°C
Zinc	850°C

Die casting

Die casting is the term used for the processes of casting metals with a low melting point into alloy steel dies (or moulds). It is known as a permanent mould process, and the molten metal either enters the mould under the action of gravity or is forced into the mould under pressure.

The alloys cast in this way are generally zinc, aluminium and magnesium-based. Their low melting temperatures make them particularly useful for large-scale production (Table 13).

The processes involved in die casting vary due to the amount of pressure/force applied to the molten metal as it enters the mould. In general, the higher the force applied the quicker the process and the finer the detail being produced.

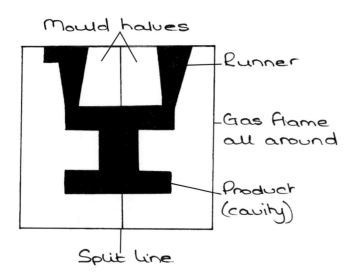

Gravity die casting

Mould halves

Runner

Gas flame all around

Product (cavity)

Split line

Gravity die casting

In this process the molten metal is poured into the dies through runners, in a similar way to that seen in sand casting. The process uses the force of gravity to ensure the molten metal reaches all parts of the metal mould.

- The dies are made from alloy steel and are split to allow for removal of the completed product.
- Gas rings around the outside of the die keep the mould heated, ensuring even cooling of the cast metal.
- Fluxes are also used to prevent oxidation of the metal as it is being cast.

As the name applies, gravity is the only force applied to the molten metal as it enters the die and makes contact with all parts of the die. Therefore products manufactured in this way tend to be large with generally simple shapes. Examples of products include car and motorbike wheels.

Alloy wheels for cars and motorbikes are produced by die casting

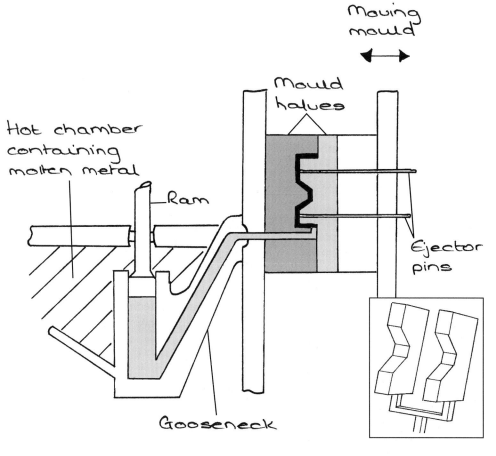

Hot chamber high pressure die casting

Pressure die casting

Die-casting processes can also use high or low pressures to force the molten material into the die. The additional pressure is required to ensure that the molten metal reaches all parts of the more intricate dies.

High pressure die casting uses a hydraulic ram to force the material into the die.

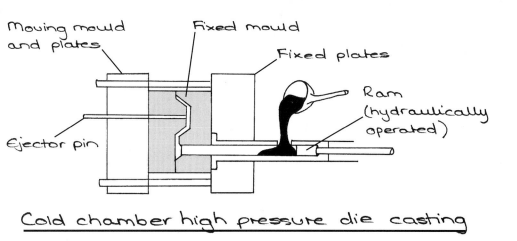

Cold chamber high pressure die casting

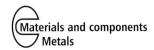
In this process molten metal is poured into a cylinder, either from a crucible or from a ladle. The hydraulic ram then forces the molten metal into the closed dies.

The dies are water cooled, resulting in rapid cooling of the product. As the dies open, ejector pins push the product out of the die.

In the hot-chamber high-pressure die-casting process the molten metal is also forced into the dies by the use of a hydraulic ram but, in this case, the ram is fed directly from the reservoir of molten metal.

Typical products of high-pressure die casting include small, highly detailed components. Examples include components for lock mechanisms for uPVC sliding doors.

Industrial die casting

A development of hot-chamber die casting is the multi-slide die-casting process. Traditional die-casting processes use just two halves of a die to form the shape – making it difficult to produce components with very complex three-dimensional shapes.

By using four (or more) slides, complex three-dimensional shapes can be achieved. Each of the components of the die is secured to one of the slides and contains either a cavity (external) or core (internal) shape which, when closed together with the other dies, will form the correct shape for the product. Each of the slides moves independently of the other for opening and closing; this is controlled by a computer and operated by pneumatics. Mechanical locking mechanisms hold the dies together while the material is being injected.

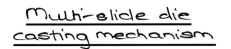
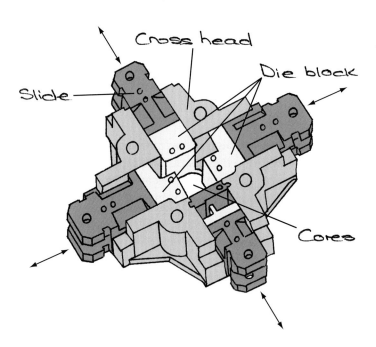

Multi-slide die casting is used for the rapid manufacture of small zinc and magnesium components. Products include door locks and sensor casings for cars and the internal components of domestic electrical sockets.

Further reading

A useful website with examples of die-casting products and the processes involved is:

- **www.dynacast.com.**

Advantages of die casting

From the products' point of view there are a number of advantages of die casting over sand casting.

- **Finish:** the surface finish of a die cast product is superior to that of sand casting; it's as smooth as the finish of the die surface.
- **Accuracy:** the shape of the die determines the shape of the product, therefore the accuracy of size and detail are as required.
- **Quality of the material:** die-cast products tend to be better from a material point of view due to the effects on the material structure of rapid cooling.
- **Scale of production:** rapid cooling of the components (<1 second per cycle) makes high-pressure die casting suitable for large-scale production – necessary to cover the cost of dies and for the manufacturer to make a profit.
- **Energy:** alloys with a low melting point require less heat to melt resulting in lower energy costs.

Investment casting

Despite investment casting being used generally for the production of casting materials with a high melting point, it is an extremely old process.

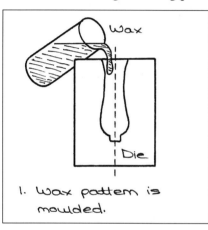

1. Wax pattern is moulded.

2. Wax runner and riser attached. Sprayed with clay.

3. Fired in kiln this bakes the clay hard and removes the wax.

4. Molten metal poured in until it appears at the riser.

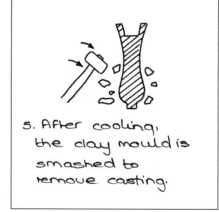

5. After cooling, the clay mould is smashed to remove casting.

6. Heat-treat to obtain desired mechanical properties. Runner and riser removed.

Stages of the process:

Step 1 A wax pattern is produced to a high degree of accuracy.

Step 2 This is then coated in high-temperature ceramic material, by dipping the wax pattern into the ceramic slip. When a sufficient thickness of ceramic material is achieved, it is left to dry.

Step 3 Once dry it can be fired in a kiln. This will of course cause the wax pattern to melt – hence the alternative term for this process 'lost wax casting' – leaving the cavity to be cast into.

Step 4 When the ceramic mould has cooled, the molten metal is poured in. This is generally done using gravity to help fill the mould.

Step 5 When the cast has cooled, the ceramic mould is broken open – thereby destroying it – leaving only the cast product.

Typical products of investment casting are:
- turbine blades for jet engines;
- tools and dies for a variety of applications;
- motorcycle steering head components;
- valves and controls for the food industry.

Advantages and disadvantages of investment casting

Advantages
- Good finishes can be obtained along with a fair degree of accuracy.
- Complicated shapes that cannot be produced by other casting processes can be made.
- Complicated shapes can be produced in materials that cannot be machined.
- There is no split line showing on the product.

Disadvantages
- The cost is very high.
- Size of components is limited by weight.

Sintering

Sintering is the process used in the manufacture of materials that are difficult to process in any other way.

The process of sintering relies on the materials being crushed into a powder. The powder is compacted into a die, which will eventually give the product its final shape. The compacted shape is then heated to promote bonding between the particles of material.

Typical products manufactured in this way are cutting tool tips and hard magnetic products made from cobalt.

Sintering – or powder processing as it is sometimes known – is also used in the ceramics industry. The process is exactly the same as for cermets and metals. Clay powder with low moisture content is pressed into shape in a hydraulic press. The clay is still in the green state, but can support its own shape until fired in the kiln where the particles of clay powder become bonded together.

Sketch showing product being preformed by compaction prior to sintering

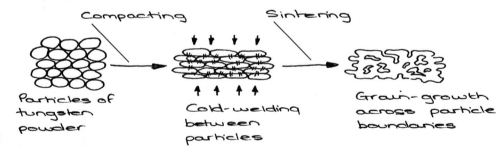

Sintering

Advantage of sintering
- Sintering is an appropriate process for materials that are difficult to process in any other way.

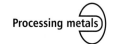

Task

Identify three products where manufacturing has involved sintering. Explain why sintering is appropriate in each case.

Forging

Forging processes can be carried out either by hand or machine. Most forging processes are carried out while the metal is hot; this avoids the risk of work hardening and also requires less energy to achieve the required result.

Basic hand processes are carried out with the use of hammers, swages and anvils. Larger forces can be achieved by the use of mechanical hammers. Processes include: bending; drawing down; punching and drifting; twisting and scrolling; and drop forging.

Bending

A bend is produced in the piece being worked; the bend can be either sharp or gradual. A more gradual bend can be achieved with the material cold, while a sharp bend will require the metal to be hot.

Drawing down

This process reduces the thickness of the material but, unlike the drawing process, which stretches the material by putting it under tension, the metal is hammered into a thinner section. This usually results in increasing the length of the piece being worked.

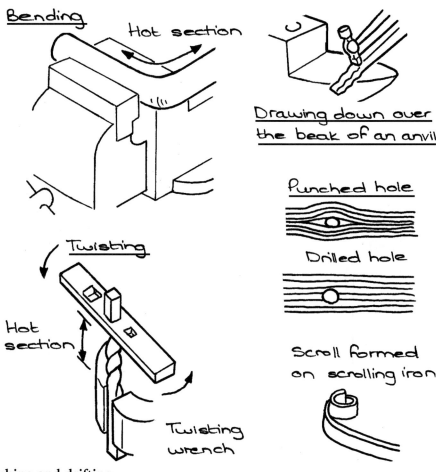

Sketches of various hand-forging processes

Punching and drifting

Punching is achieved by hammering a spiked tool into the piece being worked, while a drift is used in a similar manner to tidy up the hole that has been produced. Holes can be produced in any shape: it depends on the shape of the punches and drifts.

A horseshoe being shaped by forging

Twisting and scrolling

These two processes can be carried out with the metal cold or hot – the result will depend on the metal being forged.

All of these processes require manual labour and a high degree of skill, resulting in its suitability for relatively small numbers only.

Products made in this way include wrought iron gates, horseshoes, and stirrup irons for riding.

Drop forging

Drop forging is used where larger numbers of similarly shaped objects are required, for example, spanners and hip replacement joints. Drop forging is a refining process, for example, a piece that has been drawn down to a rough shape will be placed between a pair of drop forging dies.

One half – the upper half – is attached to a vertical sliding hammer. Very large forces are exerted onto the metal blank between the die halves, forcing it into the shape of the dies. It is usual that the component being produced will pass through a number of dies before the final shape is achieved.

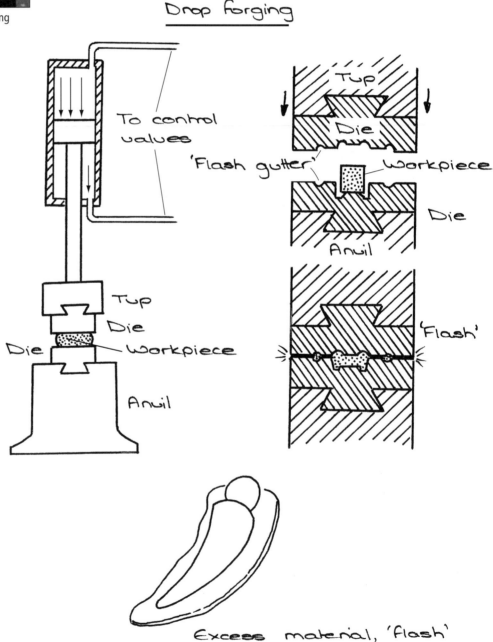

Sketch of drop forging including final product

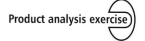

Cast iron spanner

Taking a cast iron spanner as an example, it is clear to see that if sufficient force is exerted then there is a good probability that the jaw or handle will fracture.

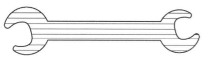

Rolled steel spanner

Materials that have been cut from sheet, on the other hand, have a parallel grain orientation due to the rolling process forming the sheet. Therefore, a rolled steel spanner will be stronger than a cast iron spanner.

Forged spanner

A spanner that has been drop forged, however, will have its grain refined and orientated resulting in a tough, strong product/component.

Further reading

For information on a full range of manufacturing processes, including milling, turning, casting, pressing and several high-tech machining processes, visit:

- **www.ee.washington.edu**

PRODUCT ANALYSIS EXERCISE: *ferrous metals*

Stainless steel saucepan

1. The saucepan pictured opposite is made from stainless steel. Explain why this material is suitable.

2. The saucepan is made by blanking and spinning. Use notes and diagrams to explain these proccesses.

3. The handle has been formed from sheet stainless steel and spot welded in place. Use notes and diagrams to explain how spot welding works.

4. Describe the health and safety measures manufacturers would take in each of the following processes:
   (a) blanking;
   (b) press-forming;
   (c) spot welding.

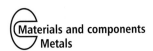
PRODUCT ANALYSIS EXERCISE: *ferrous metals*

Cast iron saucepan

1. Explain why cast iron is a suitable material for the manufacture of a saucepan.

2. Using notes and diagrams, explain how the body of the saucepan would be cast.

3. The lid handle uses a thermosetting plastic. Give an example of a suitable thermosetting plastic and explain why it is appropriate for the handle.

4. Give an example of an appropriate finish for the saucepan. Explain how this is applied.

5. Evaluate the saucepan in terms of the following:
   (a) function;
   (b) aesthetics;
   (c) ergonomics.

PRODUCT ANALYSIS EXERCISE: *alloys*

Pipe fittings, taps and valves

1. Alloys are materials made up of a mixture of two or more metals. For the garden tap here, name the alloy and list its constituent parts.

2. Explain why the alloy named above is suitable for such products.

3. The garden tap shown has been pressure die cast. Use notes and diagrams to describe this process.

4. Explain why casting is used to make such products.

5. Explain the quality control checks manufacturers would make on such products.

PRODUCT ANALYSIS EXERCISE: *non-ferrous metals*

Aluminium drinks can

1. The drinks can is made from aluminium. Explain why aluminium is a suitable material for drinks cans.

2. The main body of the drinks can has been deep drawn. Use notes and diagrams to explain how this process works.

3. The base of the can is concave. Explain why drinks cans are designed and manufactured this way.

4. Some drinks cans may be made with a steel body and an aluminium top. Give reasons why manufacturers would do this.

PRODUCT ANALYSIS EXERCISE: *non-ferrous metals*

Copper pipe

1. The pipes pictured here are made from copper. Explain why this material is suitable for the manufacture of pipes.

2. Pipes are usually made using a process known as 'extrusion'. Use notes and diagrams to explain how this process works.

3. Copper can be joined by soft soldering. Use notes and diagrams to explain this process.

4. Explain what health and safety precautions should be taken when soldering.

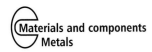

Exam questions

AS exam question

1. (a) For each of the following metals, explain why they are suitable for each application.

(i)	Stainless steel	Saucepans
(ii)	Aluminium alloy	Car engine parts
(iii)	Copper	Electrical wire

[3 × 6]

(b) Describe how mild steel could be finished to prevent corrosion. [5]

(c) Describe how aluminium can be finished to enhance its aesthetic appeal. [5]

A2 exam question

1. The physical and mechanical properties of some metals can be altered by heat treatment methods.

(a) With reference to **two** different metals that you are familiar with, describe how each metal is heat-treated to achieve a desired property. [2 × 6]

(b) For a specific product that you are familiar with, describe how specific metals are used in its construction in order to enhance the following:

(i) performance;

(ii) aesthetics. [12]

Woods

Introduction to woods

Woods have been used as both a structural material and as a decorative material for thousands of years, to provide shelter, furniture and personal decoration.

Woods are probably the oldest known natural material used by humans. Today woods are still used extensively for a variety of applications: from outdoor shelters to fine pieces of furniture, not forgetting, of course, paper making. However, there is a major challenge to preserve resources – in particular the slower growing hardwood trees.

Examples of products made from wood

At AS level

As an AS level student you should become familiar with the two main types of this naturally occurring material, and gain an understanding of the material's main properties and characteristics. A good knowledge of the different species together with examples of typical uses should be developed, as should an understanding of the role of man-made boards in the manufacture of timber-based products.

At A2 level

A2 students should build on their knowledge and understanding. This should include a thorough understanding of the use of woods, typical products and why that material is used for that product. A good understanding of the main disadvantages of using woods, e.g. rotting and attack by animals and insects, should be developed.

Sources of woods

Woods are natural materials and can be found all over the world with different species found in different areas. Approximately 80% of UK wood needs is supplied by other countries.

Types of wood

There are two basic types of tree, namely hardwoods and softwoods (see the diagram overleaf). The difference between the two types is botanical, in as much as hardwoods are generally deciduous broad-leaved species while softwoods are generally evergreen.

Examples of hardwoods grown in the UK include oak, ash, beech, sycamore and willow; while examples of imported tropical hardwoods include teak, cedar and mahogany.

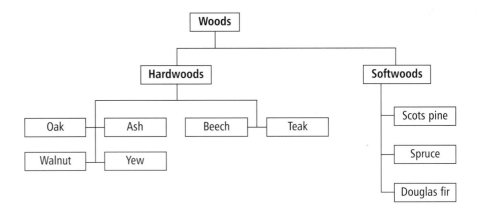

There are a smaller number of usable softwoods than there are usable hardwoods. Softwoods such as larch, spruce, Scots pine and Douglas fir are grown in forests and plantations in the UK, but approximately 90% of the UK softwood needs are supplied by countries like Norway and Sweden.

> **Task**
>
> Identify three products found in the home or school that are made from woods. Investigate the material used and give possible reasons for the choice of material.

The structure of woods

- All woods are fibrous with the fibres (or grain) growing along the length of the trunk or branch.
- These fibres consist of cells (tracheids) of, mainly, cellulose supported by lignin resin.
- Approximately 55% of the tree is cellulose, while 28% lignin resin holds it all together. The remainder is made up of carbohydrates, like sugars. Woods can therefore be thought of as a natural composite.

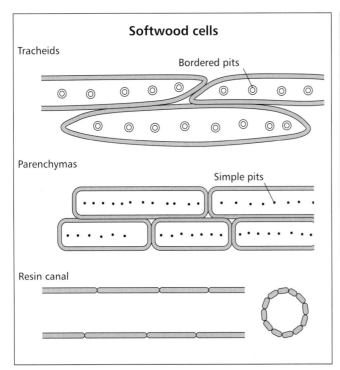

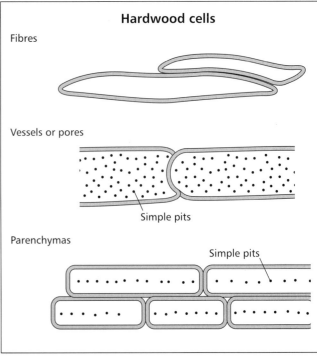

Cellular structure of softwoods and hardwoods

Strength in woods

Being fibrous materials, woods have their greatest strength in the direction of the fibres. In other words, if a length of timber were to be put under tension along the grain it would be able to support a far greater load than if it were put under tension across the grain – which would result in the timber splitting at a much lower loading.

1. Along the grain

2. Across the grain

Resulting split

Effects of putting timber under tension

Defects in woods

Woods are natural materials and therefore are not as consistent in structure as, for example, polymers.

Knots

The main difficulty with woods is that they can contain defects such as knots – these are where branches have begun to grow out of the trunk of the tree. Knots can weaken the structure of the material as well as produce an irregular grain.

✱ Conversion

'Conversion' is the term used when sawing a tree trunk into usable pieces of timber. There are two basic forms of conversion, namely (i) slab sawn and (ii) quarter sawn (of which there are a variety of methods).

As well as cutting the timber into usable shapes, the choice of conversion method can also enhance the grain and help make the material more stable.

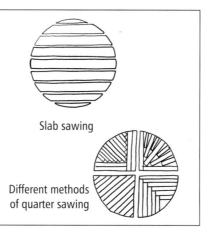

Slab sawing

Different methods of quarter sawing

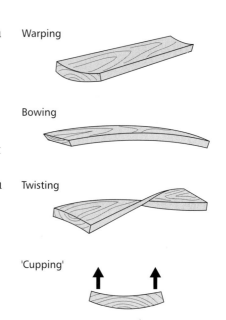

Example of defects

Splits

Other defects include splits that occur when the material is drying, or from natural events such as a lightning strike, producing hairline cracks across the grain (more often found in imported hardwoods from Africa).

Shrinkage

Shrinkage occurs as the material is dried out and loses moisture during seasoning. Moisture plays a large part in the use of timbers. For example, if a timber product is to be used outside, then if it were dried to a 5% moisture content it would very quickly absorb moisture resulting in the material being more prone to rotting. Conversely, a timber product with a higher than 5% moisture content for indoor use would quite quickly dry out leaving it with unsightly splits and evidence of shrinkage.

Warping, bowing, twisting and 'cupping' are deformations of the material due to uneven shrinkage.

Dry and wet rot, insect attack

Other defects found in woods include attacks by fungus (dry rot), wet rot and attack by insects.

- Dry rot reduces the wood to a dry, powdery consistency resulting in little strength.

- Wet rot occurs where there are both damp and dry conditions. This alternating cycle breaks the material down by decomposing it.

- Insects in the form of, for example, woodworm (seen in softwoods) and deathwatch beetle (seen in hardwoods) attack the material by laying their eggs in the pores or in crevices. These eggs hatch into larvae that then proceed to tunnel their way through the cellulose structure of the material, until they emerge through flight holes as beetles.

Warping

Bowing

Twisting

'Cupping'

Defects due to drying

Seasoning woods

Seasoning is, in effect, a controlled drying of timber. This can be achieved either through natural seasoning or by kiln drying. Whichever method is used, the moisture content of the material must be below 20%. The ideal is that the moisture content of the timber is the same as that of its surroundings – known as the Equilibrium Moisture Content (EMC).

The benefits of seasoning

Here are the benefits of seasoning.

- It increases the strength and stability of the timber.
- The reduced moisture content reduces the risk of timber causing corrosion to surrounding metalwork.
- It makes the timber less prone to rot and decay.

Old timber-framed house

Green timber

Not all timbers are seasoned thoroughly. The term 'green timber' refers to woods that have been cut down, converted and stored for up to 12 months. An example of the use of green timber is in the construction of timber-framed houses. These are usually architect-designed one-off structures. Using green timber in this way has the benefits of the aesthetics of the grain of the wood, along with drying cracks that give the material character.

Veneers, laminates and composites

Veneers

'Veneer' is the term given to a thin layer of wood that has been shaved off the trunk of a tree. Hardwoods are the usual materials to make up veneers, because they tend to be more decorative and durable than softwoods.

Veneers are used in sheet form to provide a more decorative surface to inferior quality woods. For example, a veneer of yew could be applied to a piece of chipboard that would be used in the manufacture of a hi-fi or television cabinet (see Table 14).

Product finished with a wood veneer

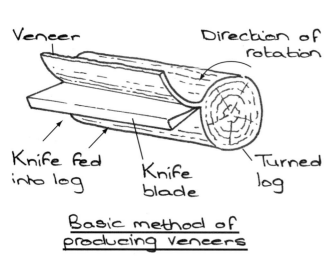

Veneer

Direction of rotation

Knife fed into log

Knife blade

Turned log

Basic method of producing veneers

The use of veneers can reduce the cost of a product by enabling the manufacturer to use lower-cost materials for the main structure, and then finishing in a veneer. This reduces the need to purchase solid timber, which would be too expensive for mass-market products.

Table 14: Veneers and base materials

Suitable veneers	Suitable base materials
Beech	Medium Density Fibre board (MDF)
Oak	Chipboard
Ash	Block board
Walnut (including burr walnut)	Plywood
Yew	

Laminates

A laminate is a material that has been placed in layers with the same or other materials. Examples of laminates include the Formica surfaces for worktops and laminate flooring products.

Formica surfaces

Formica worktops are made up of layers of paper in a melamine resin – the top layer of paper gives the required decorative finish while a layer of hardwearing, heat-resistant polymer is the final layer. The base material for the whole worktop would be a particle board such as chipboard.

Laminate flooring

Laminate flooring is made up of layers of printed materials in a resin on a supporting material, such as MDF. The final top layer of the flooring is a very hardwearing thermosetting polymer.

Plywood

The previous two examples show how polymer-based laminates can be used with wood products. Plywood, on the other hand, is a good example of the use of laminated veneers to produce large sheets of stable material with a natural wood finish.

Applying a laminate

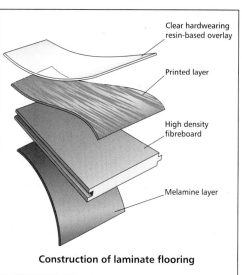

Clear hardwearing resin-based overlay

Printed layer

High density fibreboard

Melamine layer

Construction of laminate flooring

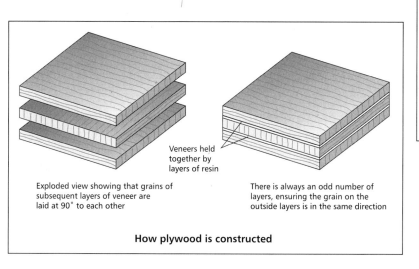

Veneers held together by layers of resin

Exploded view showing that grains of subsequent layers of veneer are laid at 90° to each other

There is always an odd number of layers, ensuring the grain on the outside layers is in the same direction

How plywood is constructed

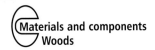
Layers of hardwood veneer are bonded together. The grain direction in each additional layer is laid at 90° to the previous layer. The result is a sheet material that is very stable against warping.

Task

Plywood can be made up of a number of layers resulting in a variety of thicknesses. However, there is always an odd number of layers. Explain why this is the case.

Product made from laminated wood

Laminated wood products

As we have seen, a laminate is made up of layers of material bonded together. The advantage of this is that it produces a very stable material. We can produce products from laminated timber. For example, the seat of the dining chair shown here has been produced as a laminated structure. Layers of hardwood veneers have been placed in between two halves of a mould. In between each veneer is a layer of resin. As the mould halves close, it presses together all the layers of veneer and forms the shape of the seat. It may be necessary to soak the veneers prior to moulding to make them more supple.

The range of laminated wood products includes CD racks and furniture, along with structural items such as moulded beam trusses for sports venues.

Wood-based composites (man-made boards)

We have discussed composites earlier in this book (pp.22–31) but another look, particularly at wood-based composites, would be appropriate here.

In fact, timber can be thought of as a natural composite due to the structure of the material containing cells that are supported by natural resins.

In the same way a number of wood-products – the term given to those materials made from wood residues or smaller pieces – are also composites. See Table 15.

Table 15: Wood products

Man-made board	Made up of:
Plywood	Layers of veneers and resins; always an odd number of veneers
Block board	Strips of wood bonded together with a veneered surface
Chipboard	Fine chips of woods mixed with resins
MDF	Very fine wood fibres mixed with resins
Hardboard	As MDF – can be impregnated with oil
Sterling board	Shavings of wood compressed into resins

Task

Investigate how man-made boards have been used in the construction of a house and in the manufacture of furniture in various rooms of a house. Identify the types of man-made board used and give reasons for their use, including details of any finishes that have been applied.

Further reading

Two useful websites covering a range of aspects of working with woods are:

- **www.woodmachining** for articles on wood machining, software, cutting tools and safety;

- **www.woodweb.com** for an extensive collection of articles from the wood industry, including materials and components, machining, joining and finishing.

Natural timber and manufactured boards

Natural timber and manufactured boards are extremely popular materials with designers and manufacturers. Such materials not only have excellent functional and aesthetic properties, but are also more sustainable than other materials. One company that uses timber and manufactured boards extensively is IKEA.

IKEA prides itself on making good-quality contemporary products at an affordable price, with minimum impact on the environment. Because IKEA and its licensed suppliers make huge numbers of wooden products, it is very important that the timber they use is from sustainable sources.

Where IKEA uses natural timbers such as pine or birch, it will ensure that the timber is harvested from managed forests. In such woodland, as one tree is felled, it is replaced with several saplings to ensure that timber is available for future generations. In addition to this, IKEA generally uses timbers that are fast growing, maturing at, say, 30–40 years (as opposed to some hardwoods, such as oak, that take over 80 years to grow to maturity). Therefore the timber it uses can be replaced in a relatively short time period.

One product that is made from timbers is the aspen CD rack. Aspen is a timber that grows rather like a weed. It is widespread, grows rapidly and, until recently, has not had any particular commercial value. IKEA makes veneers from aspen to laminate into products such as the CD rack pictured below.

Some designers choose to use timber for its aesthetic properties. The garden designer Diarmuid Gavin, famous for his contemporary garden structures, often uses hardwoods such as oak for its aesthetic effect in, for example, a garden building such as a summerhouse or gazebo. If green oak is used, in addition to its being easier to use, as it is soft when wet, the timber will tend to split as it dries out. This can add to the overall aesthetic effect and give the timber product a more natural appearance, which is desirable in a garden. Oak has the added advantage that it does not need to be finished in order to preserve it. However, finishes such as combined stains and preservatives may be added to give the timber a colour.

Aspen CD rack

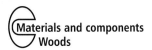
PRODUCT ANALYSIS EXERCISE: *woods and manufactured boards*

Laminated dining chair

Study the photograph of the dining chair.

1. The chair is made from laminated beech veneers. Explain why beech is suitable, making reference to the following requirements:
   (a) aesthetic;
   (b) functional;
   (c) manufacturing.

2. Use notes and diagrams to sketch how the veneers are laminated into the shape of the seat.

3. Sketch how the seat could be joined to the legs.

4. The seat is finished with a polyurethane varnish. Explain why this is suitable and how it would be applied.

5. The table top pictured is made from MDF. Explain why this material is often used in such furniture.

6. The table top is finished with a melamine formaldehyde laminate. Explain why this is used.

PRODUCT ANALYSIS EXERCISE: *woods and manufactured boards*

Kitchen utensils

1. (a) Compare and contrast the two sets of kitchen utensils. Make reference to the materials used, their suitability for their purpose and their methods of manufacture.

   (b) Use notes and diagrams to show how the handle of the stainless steel utensils could have been joined to the rest of the utensils.

   (c) The kitchen utensils could have been made from a polymer instead. Name a suitable polymer and explain why this would be suitable.

Exam questions

AS exam question

1. (a) For each of the following timbers, explain why they are suitable for the applications listed.

(i) Beech 50 mm × 50 mm section	Turned pieces, e.g. saucepan handles
(ii) Tannelised rough sawn Scots pine 50 mm × 25 mm × 10 m	Framework for construction jobs; fence/shed making
(iii) Balsa 10 mm × 10 mm × 300 mm	Model making

[3 × 6]

(b) Explain the health and safety precautions you would take when working with hardwoods. [5]

(c) Explain how a timber might be finished to enhance its aesthetic appeal. [5]

A2 exam question

1. (a) Architects are making increased use of timber as a construction material for contemporary buildings. With reference to factors such as sustainability, ease of use and aesthetics, explain why timber is being used in contemporary building design. [12]

(b) Explain how timber used in the construction of buildings can be finished to preserve it. [12]

Glass

Glass as a material

There are a wide range of materials that can be classed as glass, but for the purposes of this book glass is essentially the material used for glazing windows, manufacturing bottles and so on.

Manufacturing with glass

Lime-soda glass is made from a mixture of sand, lime and sodium carbonate that is heated to 1500 °C in a large furnace. The resulting molten material is then tapped off to form a continuous flow that can then be 'floated' on a tank of molten tin to form a glass sheet – called plate glass.

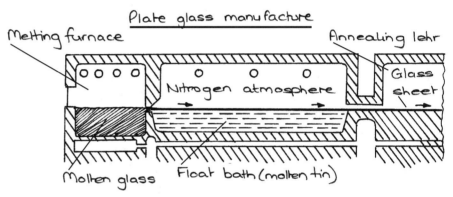

Plate glass manufacture

Melting furnace
Annealing lehr
Nitrogen atmosphere
Glass sheet
Molten glass
Float bath (molten tin)

Glass made in this way is a mixture of new materials and recycled materials called 'cullet' obtained from recycling centres. Up to 90% of cullet can be included in the furnace.

Glass blowing

Glass blowing is used to manufacture hollow objects such as bottles. Mouth blowing is restricted to the more expensive pieces of glassware. For everyday products, a more automated process is used where a 'gob' of glass is formed by a cycle of pressing and blowing into a mould.

Hand-crafted glass products

Glass blower at work

Slumping

'Slumping' is the process where glass is heated until it becomes sufficiently soft, allowing it to take the shape of a mould. Special curved glass for windows is produced in this way, as are car windscreens.

A variation of the use of slumping involves filling a ceramic mould with a mixture of clear and coloured glass fragments (cullet). This is then placed into a furnace. The glass becomes molten taking up the shape of the mould. The result of this process is a block of glass patterned by the way the coloured glass has flowed into the clear material.

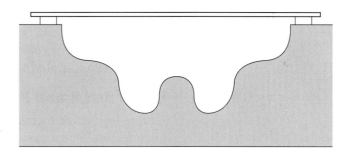

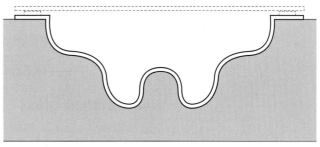

Slumping into a mould

Annealing glass

Glass, like most other materials, contracts on cooling and in doing so internal stresses can develop between the more rapidly cooled outer surfaces and the core of the material. To remove these stresses, and so reduce the risk of the material cracking, the glass product is cooled very slowly in an oven called a 'lehr'.

General properties of glass

Glass is extremely brittle. Mechanical shock will cause the material to break, e.g.
- impacts such as dropping;
- a cricket ball, for example, hitting a pane of glass;
- thermal shock, e.g. placing an already hot glass into ice or cold water (or vice versa).

Glass is much stronger in compression than in tension and has a good chemical resistance.

Toughened glass

As in most groups of materials, the properties of glass can be modified. Glass used in the construction industry, for example, can be toughened by heating to 400 °C followed by rapid cooling with air blasts. This produces compressive forces on the two outer surfaces on the glass, while the internal core of the material remains in tension. In order for this glass to break, an external force must overcome the internal stresses. No further processing of toughened glass is possible, since any cutting or damage to the outer surface will cause the glass to

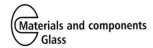
shatter. Glass that has been processed in this way is used for vehicle windows and glass doors.

Coloured glass

Glass can be coloured by adding oxides of metals such as iron, copper and nickel. Examples of coloured glass products are stained glass windows, wine and beer bottles and tinted windscreens for cars.

Lead crystal and Pyrex

Adding lead oxide increases the way the material reflects light and is used in the manufacture of 'cut glass' wine glasses, decanters and vases.

'Pyrex' is the manufacturer's name for a range of heat-resistant products made from boro-silicate glass. In this material, boron oxide is added to the glass mixture. Its resistance to high temperatures makes Pyrex ideal for cookware applications and for laboratory equipment.

Other oxides can be added to produce specialised glasses for specialist applications, like the windows used in spacecraft.

Laminated glass

Laminated glass is a very tough composite material made up of two thin sheets of plate glass between which is sandwiched a sheet of tough clear polymer. If the glass is broken, the polymer holds the fragments together. This makes the material suitable for security applications.

Glass as a thermal insulator

Heat from the Sun enters through the glazing.

Double glazing

Inside building

How emissivity coating on this surface prevents heat from radiating outwards.

Sketch showing the benefits of coated glass

Pilkingtons are probably the largest manufacturer of glass in the UK. It has developed a glass that acts as an insulator to heat. This is known as 'K' glass and allows heat (via sunlight) into a building while reducing the amount of heat that escapes.

Self-cleaning glass

Self-cleaning windows are a new development from the glass industry. A special coating is applied to the outer glass surface, which is virtually invisible. The coating:

• prevents droplets settling on the surface by forcing the water to spread out into a sheet, thereby reducing the formation of spot marks;

• interacts with ultraviolet light to break down organic dirt (e.g. finger marks, dust, pollen) into simpler compounds that can then be washed away by rain.

This new type of glass is ideal for inaccessible windows, e.g. windows high up in tower blocks.

Advantages of replacing glass with polymers

If you think about all the products that are or have been produced from glass, it is more than likely that polymers have been or could be used to replace it. The exception comes when the products used involve high temperatures, i.e. above 200 °C.

Having said that, some ready-made meals available from supermarkets can be placed in a conventional oven at up to 200 °C while the contents are being cooked. These containers are usually made from a thermoplastic polymer, such as PET, or a food quality PVC. This is possible due to developments in polymer technology.

A further example of where polymers have replaced glass is in the manufacture of bottles. Think about all those products that could be contained in a glass bottle. Examples include: foodstuffs like jams and sauces, wines and spirits, medicines, cleaning fluids, etc. A large range of these is now contained in plastic bottles.

Polymer-based food container

Advantages of using polymers instead of glass

- They are lightweight (for transportation).
- They have a low melting point (recycling).
- They can be impervious to gases (carbonated drinks in PET).
- They can be squashed without breaking.
- Screw tops can be used.

Advantages and disadvantages of using glass instead of plastics

Advantages

- Glass is a more rigid material than a polymer, especially thermoplastic polymers.
- Glass is more scratch-resistant than most polymers.
- Generally, glass has a greater clarity where transparency is required.
- Glass is not affected by heat in the same way as most polymers.
- Glass can be used where heat resistance is required, e.g. in kitchenware.
- Glass gives a sense of quality to a product due to its weight, texture and light-handling qualities.

Disadvantages

- Glass is heavy (increased transportation costs).
- Glass has a high melting point (more energy needed for recycling).

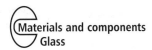

Exam questions

AS exam question

1. (a) Explain why glass is a suitable material for the manufacture of drinks bottles. [6]

   (b) Name an alternative material that is suitable for the manufacture of drinks bottles. [2]

   (c) Explain why this is a suitable alternative material. [6]

   (d) Use notes and diagrams to show how a bottle would be manufactured using this material. [9]

   (e) Explain what is meant by the term 'biodegradable material'. [3]

A2 exam questions

1. Advancements in materials and manufacturing technology have significantly enhanced the performance of products. For **two** of the following materials, explain in detail what advancements have been made and describe how they have improved the performance of specific products:

   ● K-glass;

   ● shape memory alloy;

   ● carbon-fibre reinforced plastic. [2 × 12]

2. For each of the following glass materials, describe a suitable application and explain why they are suitable:

   (a) K-glass;

   (b) laminated glass;

   (c) toughened glass;

   (d) recycled glass. [4 × 6]

Ceramics

Introduction to ceramics

We tend to think of ceramics as those materials used to produce crockery (cups, saucers, plates, etc.) and ornamental and decorative products (vases, figurines, tiles, etc.). There is, however, a wide range of ceramic materials that include engineering applications, such as the tiles for the NASA space shuttle.

The space shuttle

Most of the more common, everyday ceramics products are clay-based materials. These include a range of clays with different additives, providing products with a range of different finishes. Some clays can be gritty in nature, others can provide high-strength products for use in hotels, or cheap earthenware for everyday domestic use.

At the top end of the market is fine bone china. This type of clay contains finely ground animal bone giving the fired material its translucent quality.

Engineering ceramics

Most ceramic products are made from clay. Engineering products include house bricks, engineering bricks (once used in the construction of sewers – not yet superseded by concretes or plastics), electrical insulators for pylons, etc. There are an increasing number of high-temperature applications where metal-oxide based ceramics are being used. Table 16 shows the materials used, their maximum working temperature and potential applications.

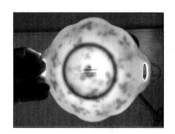

Fine bone china showing its translucent qualities

Table 16: Metal-oxide based ceramics

Substance	Melting point	Common uses
Alumina	2050 °C	Spark plugs, crucibles, cutting tools
Beryllia	2350 °C	Crucibles for high-temperature nuclear reactors
Magnesia	2800 °C	Furnace linings
Zirconia	2690 °C	Liners for rockets, wall insulation for high-temperature furnaces

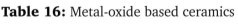

Useful features of engineering ceramics

- Ability to withstand high-temperatures without distortion;
- strength and rigidity at high-temperatures;
- freedom from 'creep' (essentially grain growth that increases the size of the product);
- hardness and resistance to wear.

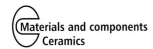

Decorating ceramics

Most domestic ceramic products have been decorated in some way, whether it be a plain coloured mug or a highly decorated dinner plate. There are a number of methods that can be employed, including:

- coloured glazes;
- under glaze and on-glaze decals;
- total transfer methods, employing silkscreen printing.

Glazes

A glaze is initially slurry made up of very finely ground glass particles held in an aqueous (water-based) solution. Ceramic products such as domestic crockery or sanitary ware are either sprayed with or dipped into the glaze.

Once dried, the product is fired in a kiln leaving a high-gloss, non-porous finish that is resistant to stains, detergents and acids.

Colours can be added to the glaze to give, say, a blue mug or green pedestal sink.

Under glaze and on-glaze patterns

As the name implies, these patterns are applied at different stages of the manufacturing process. Products will have either an under glaze pattern or an on-glaze pattern applied during their manufacture.

Under glaze patterns are put on to the product before the clear glaze is applied, while the on-glaze patterns are fitted to the product after glazing and firing. The disadvantage of using on-glaze patterns is that an additional firing stage is required to secure the pattern to the glaze.

The method of applying the pattern is the same in both cases. A decal is produced by printing the required pattern onto gummed paper. By soaking the decal in water, the pattern can be removed from the paper and lifted as one piece to be fitted onto the fired product. It's a bit like applying transfers to a plastic kit. The process is still carried out largely by hand, followed by a light sponging to remove any trapped air. The product is ready then for either glazing or firing.

Total transfer printing

This method of applying colour to ceramic products has been developed over the last 10 years or more, and involves the silkscreen printing of a pattern on to a flat substrate material. A flexible silicon dome-shaped form is lowered onto the substrate and, as it lifts off again, it takes the printed pattern with it. The product is placed under the silicon dome, which is lowered. When it is lifted again, it leaves behind the pattern on the ceramic product.

Up to four colours can be applied in this way onto: flatware, such as saucers, plates and shallow dishes; the outside of cups and mugs; the outside of flat products, like wall tiles. (See diagram, overleaf.)

Task

Investigate haw decorative patterns have been applied to ceramic products that can be found around the home, e.g. in the kitchen.

You should consider a variety of products, including all types of tableware.

Total transfer printing

(a)

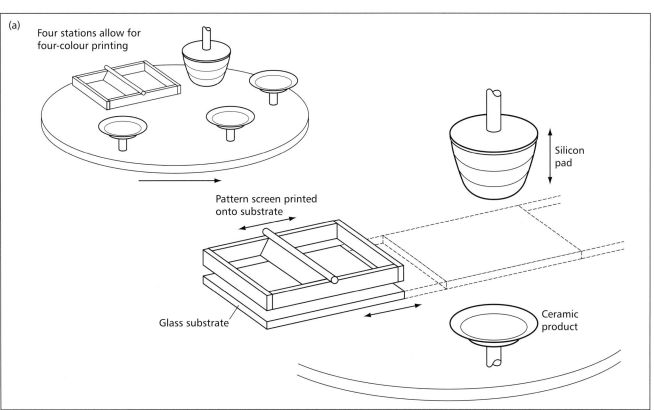

Four stations allow for
four-colour printing

Silicon
pad

Pattern screen printed
onto substrate

Glass substrate

Ceramic
product

(b)

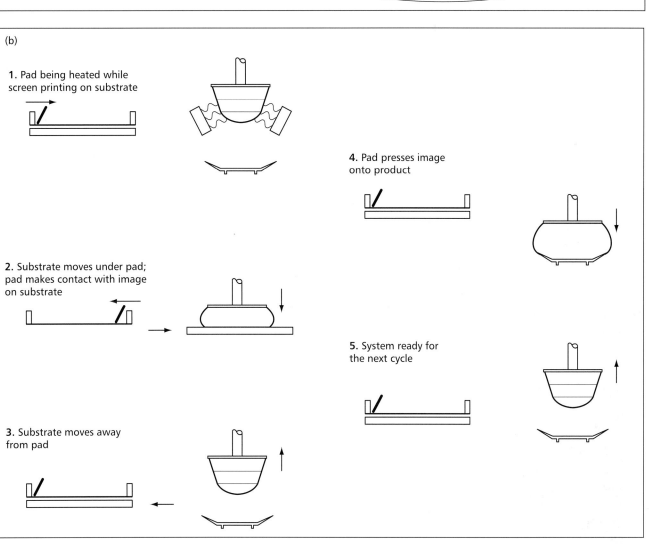

1. Pad being heated while
screen printing on substrate

2. Substrate moves under pad;
pad makes contact with image
on substrate

3. Substrate moves away
from pad

4. Pad presses image
onto product

5. System ready for
the next cycle

PRODUCT ANALYSIS EXERCISE: *ceramics*

Bone china milk jug

1. The milk jug shown here is made from fine bone china. State what is meant by the term 'fine bone china'.

2. The jug is made by slip casting. Using notes and sketches describe the process of slip casting.

3. Use notes and sketches to describe the process of glazing. Explain the benefits of glazing a ceramic product.

4. The jug is decorated with an on-glaze lithograph. Explain the term 'on-glaze decoration'.

Exam questions

AS exam question

1. (a) Describe how this ceramic jug could be made using a slip-cast method of manufacture. [10]
   (b) Describe the benefits of manufacturing such products with ceramics. [5]
   (c) Describe the disadvantages of manufacturing such items with ceramics. [5]
   (d) Describe the health and safety precautions manufacturers would take to protect employees working with ceramics. [8]

A2 exam question

1. Compare and contrast the two plates pictured below. You should make reference to the materials used, the methods of manufacture and the benefits of each. [2 × 12]

Papers and boards

Introduction to papers and boards

At AS level

As an AS level student, you should develop an understanding of the range of papers and boards available and how appropriate these materials are for different purposes.

At A2 level

A2 level students should again build on the knowledge and understanding of these materials, being prepared to offer explanations as to why and where they might be used.

Types of paper

Designers must have a mechanism by which to demonstrate ideas or final products to clients and so receive feedback as to the potential success of proposed solutions. Some papers are more appropriate for specific media and drawing techniques.

For example, cartridge papers provide a good surface for sketching and using coloured pencils for rendering. There are a variety of qualities but the heavier papers are generally more versatile, being less prone to yellowing with age and likely to be acid-free.

Watercolours can be used with cartridge papers, but to prevent 'cockling' the papers must first of all be stretched, hence special watercolour papers are available. These provide a good surface texture that will accept acrylics, gouache and pastel, as well as watercolour.

There are three main textures available in watercolour papers.

- Hot-pressed papers have a hard, smooth surface.
- Cold-pressed papers have a rougher surface, which enhances the finished image by allowing more of the colour to be absorbed.
- A third type is rougher than cold-pressed paper, having more peaks and hollows (known as tooth) on its surface.

Ingress papers are available for use with pastels, while bleed-proof papers are useful for working with markers. Pastels can also be used with coloured sugar papers giving a more textured surface than ingress paper.

Other papers are available to the designer for more formal, engineering drawings, showing details such as dimensions and construction details.

Paper properties and uses

Optical properties

The most important optical properties for papers are brightness, colour, opacity and gloss. (Brightness is the degree to which white or near-white papers and cards reflect light.)

Opacity

An opaque paper will allow little or no 'show through' of the image from the other side of the paper. This is one of the most desirable properties of writing and printing papers.

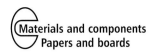
Gloss, glare, finish and smoothness

These terms are used to describe the surface characteristics of paper.
- Gloss refers to surface lustre.
- Glare refers to the way the paper reflects light.
- Finish refers to the general surface characteristics of the paper.
- Smoothness refers to the absence of surface irregularities.

Strength and durability

The strength of paper is determined by the following factors:
- strength of individual fibres;
- average length of fibre;
- strength of bonds between fibres;
- structure of the paper.

Strength falls away rapidly with the increase in moisture, due to the breakdown of inter-fibre bonding.

Tensile strength

Most papers require a certain minimum strength to withstand the production processes: including printing, embossing and folding, as well as handling.

Bending strength

The thinner the sheet, the more flexible and light it is; conversely the thicker and heavier a paper is, the more stiff it is.

Porosity

Porosity is reduced with the addition of size to the paper. Greaseproof paper is made by beating the paper, resulting in a dense sheet with very little porosity.

Manufacturing with papers and boards

Papers and boards are used widely in the manufacture of products such as packaging for flat-pack products, flowers and electrical goods, as well as products such as paper cups, newspapers and magazines. All of these products require materials that meet the functional requirements.

Cards and boards

Heavy papers can be stiff, making them useful for model making. Carton boards are usually quite smooth, with one of the surfaces being white while the other is left natural.

Cardboards are available in a range of thicknesses. They are made up of a number of layers, the middle layer being corrugated, giving the material significant strength and ability to provide protection. These materials are used in packaging and can be cut and folded from flat sheet.

Foam boards

Foam boards are a multi-layer board made up of two outer layers of card.
- The outer surfaces have a high gloss finish.
- The middle layer is foam.

Foam board has a number of uses in the design process, from mood and presentation boards to use as a modelling material. It is generally light in weight, easily cut, but difficult to bend.

A selection of foam boards

Correx boards

These boards are produced by extruding a thermoplastic to produce a sheet material that is useful for the manufacture of simple products. This material is lightweight and durable and can be bent easily in one direction only. Special fasteners can be used with correx, making it possible to manufacture products.

Printing

There are a large number of products available that contain graphics to identify and promote the company and product, provide information regarding contents of food products, weight and warnings as to whether a product is hot or heavy, copyright information and barcodes. Examples include packaging for breakfast cereals or fast-food outlets. Other products can contain graphics that are purely decorative, for example, decorated china tableware.

These products are printed using a commercial printing process. The diagram below gives an overview of printing processes available.

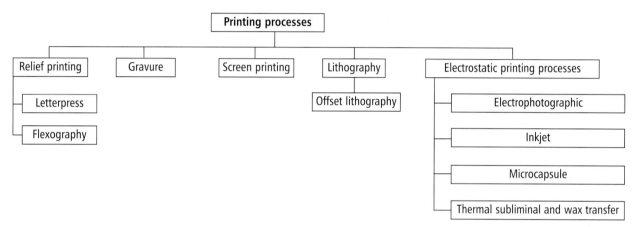

Printing processes

Commercial printing processes include:

- lithography;
- letterpress;
- flexography;
- gravure;
- screen printing.

Processes such as letterpress can only print one colour at a time; screen printing can only use solid colours: for a multi-colour image a number of screens must be used.

For processes like lithography, flexography and gravure, multiple colours can be printed in one pass through a press. This is known as process colour printing and uses four transparent inks:

- cyan (blue-green);
- magenta (red);
- yellow;
- black.

The four colours are printed one on top of another in varying amounts to produce the density and tone of colour required. Colour photographs and other artwork can be faithfully reproduced in this way. Spot colours, on the other hand, use custom-mixed inks (or Pantone colours) and are widely used in package printing, where large areas of uniform colour are common.

Letterpress

Letterpress is one of the earliest methods of printing. This process involves the use of raised letters onto which a coating of ink is deposited. By pressing the plate holding all of the letters onto the material being printed, e.g. paper, the ink will be left behind – producing the printed document. Letterpress is still used today for specialist printing, e.g. personalised wedding invitations.

Offset lithography

Offset lithography is probably the most versatile and economic of the commercial printing processes. The process uses the same four colours: **c**yan, **m**agenta, **y**ellow and blac**k** (CMYK). One-, two-, three-, four- or five-colour presses can be used, depending on the number of colours being printed. For example, black text on white paper would use a single press, whereas a modern newspaper with colour photographs would be printed using the four colours (CMYK). Five-station printing presses can be used where the fifth station prints a spot colour (a single colour other than the CMYK colours) or a varnish.

Stages of the process:

Step 1 Printing materials can be fed into the machine as a sheet (as in carton printing) or as a web (from a roll – as in newspaper printing). This process is suitable for batches of 1000 items or more at speeds of 4000 to 12 000 impressions per hour.

Step 2 Printing plates are produced from photosensitive aluminium, the image being etched onto the plate using lasers. The plates will then be fitted to the machine and a test run will be carried out, to ensure all registration marks line up and that the colour is the correct density.

Step 3 The printing process itself relies on the plate being wetted with a damping roller, while the grease-based inks will only go into those regions where required. The plate cylinder rotates onto a blanket roller, which then becomes coated with the ink, which, in turn, is transferred onto the paper.

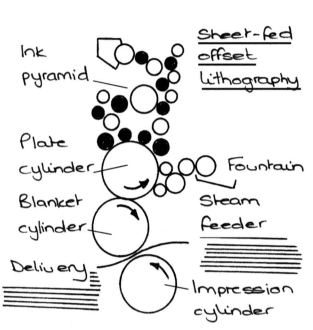

Sketch of roller arrangement for sheet-fed offset lithography

For double-sided printing, the same arrangement of rollers is positioned under the press. This type of double-sided printing is often used in very large printing presses for printing newspapers, while smaller printing presses rely on the rotation of the sheet materials to produce double-sided work.

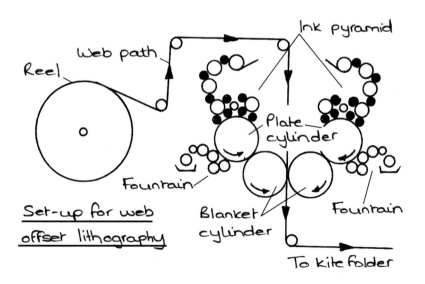

Sketch of roller arrangement for web offset lithography

Flexography

This is a relief printing process – similar to letterpress with the raised letters. In this case, the raised text and images are photo etched onto rubber material that is glued onto steel rollers.

The inks used in offset lithography are quite thick and viscous, whereas those used for flexography are thin. This allows poor-quality materials to be printed on, including waxed boards and cellophane. Primarily this process is used for food packaging – including plastic carrier bags – although magazines and paperback books can also be printed using this method.

Flexography is a good process for printing on cheap materials, and prints well on non-absorbent stock. It is expensive to set up, so long runs are usually preferred.

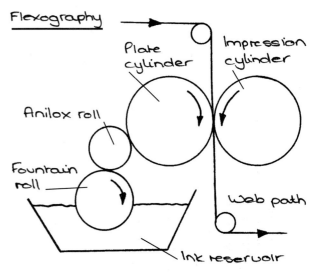

Screen printing

This printing process has a number of different applications where inks or other materials may need to be printed onto a flat surface.

Uses of screen printing
- The transfer of designs onto papers and fabrics;
- the transfer of designs onto ceramic products, e.g. tiles, plates, etc.;
- for printed circuits for electronic products, cartons, advertising boards and compact discs/DVDs.

The process
Originally called silkscreen printing, this process relies on ink being forced through a mesh. Part of this mesh is blanked off while the area of the required image is open. A rubber squeegee forces ink through the open area of the mesh onto the material being printed.

This is a versatile process that can be used to print onto curved surfaces, for example, bottles. However, only one colour can be printed at a time so for images requiring more than one colour the process will be repeated the required number of times.

This process can be manually operated or fully automatic, depending on the scale of production. For example, the printing of tiles and plates in the ceramics industry is fully automatic (see Decorating ceramics, p.74).

Advantages of screen printing
- It is able to print thick deposits of ink on uneven surfaces.
- It is also a non-impact method of printing, so wear is much reduced.

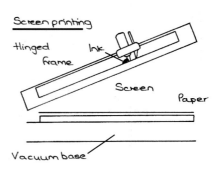

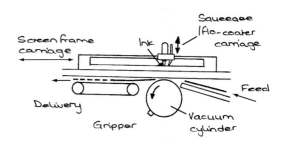

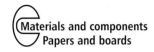
Gravure

Gravure is used to produce a range of generally high-volume materials such as postage stamps and banknotes through to catalogues and wrapping paper.

The process involves the engraving of a stainless steel cylinder with thousands of tiny holes of varying depth. These holes give the tonal variation required in the image. Ink is collected in these holes from a reservoir as the cylinder is rotated. The cylinder is wiped clean with a 'doctor blade', leaving ink only in the area of the image.

The material being printed onto is web-fed, and once printed it is folded and dried.

Gravure is a very expensive process, so is therefore reserved for very long runs. Any alteration to the design requires a new cylinder.

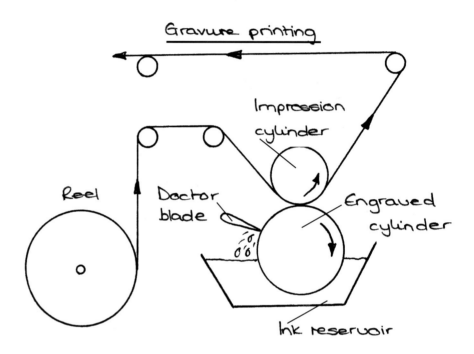

Sketch showing roller arrangement for gravure printing

Advantages of gravure
- It produces high-quality print.
- It prints at high speed.

Electrostatic printing processes

Digital printing

Digital printing processes include electro-photographic and inkjet printing.
- Electro-photographic printing involves the depositing of toner onto the substrate (e.g. paper, card). The amount of toner being deposited is controlled by varying its electrostatic properties. Dry or liquid toners can be used. These are fixed by absorption, heat or chemical reactions.
- Inkjet printing involves spraying electrostatically charged ink droplets directly onto the substrate, relying on absorption or heat to fix the inks.

Advantages of digital printing
It has a much reduced make ready time, therefore:
- It's economical for short runs of up to 1000 copies.
- It's ideal for 'on-demand' printing.

Thermal transfer printing

Thermal transfer printers apply the process colours (cyan, magenta, yellow and black) one at a time by heating the wax to leave droplets of colour on the substrate (i.e. paper or acetate sheet).

Dye sublimation printing

Dye sublimation printing is a form of thermal transfer printing, but works in a quite different way. The dyes used change from solid to gas without becoming liquid at any point. The thermal printing head varies the temperature of the dye, and so controls the amount of dye being printed onto the substrate (e.g. paper). This means that the image is continuous and is not made up of dots of colour (called dithering) as in electrostatic and thermal (wax) transfer methods.

Die cutting, creasing and folding

Products like this cardboard 'vase' used by florists are produced from a flat sheet of corrugated cardboard. The product is supplied to the florist in the form of a flat sheet, pre-cut to the required shape with fold lines already creased.

One side of the cardboard sheet has been printed with one colour only. The shape has been cut using a die cutter. The die cutter comprises two types of blade, one of which is the cutter while the other blade has a rounded edge that will produce a crease in the cardboard ready for folding.

Die cutting is used in the production of a variety of paper and card products.

Specialist machines can do folding of papers and cards automatically. These machines are programmable, giving great flexibility in the combination of folds that can be achieved.

Die cut cardboard vase

Die cutter

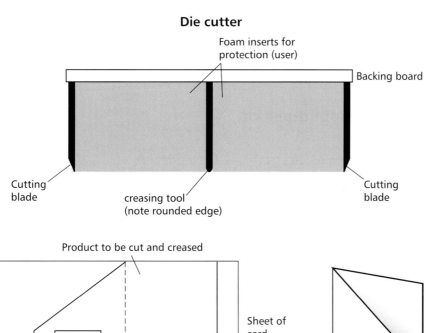

Foam inserts for protection (user)

Backing board

Cutting blade

creasing tool (note rounded edge)

Cutting blade

Product to be cut and creased

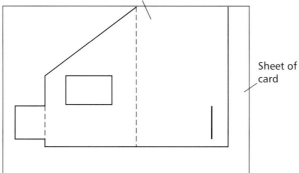

Sheet of card

A diecutter may be required to cut (where solid lines are shown) and/or crease (where dotted lines are shown)

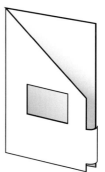

Example of a card wallet shaped and folded after being die cut

Die cutter

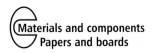

PRODUCT ANALYSIS EXERCISE: *papers and boards*

Packaging

1. Prepare a short PowerPoint presentation to cover the following details:

(a) PET and its suitability for drinks bottles;

(b) polystyrene and its suitability as a packaging material;

(c) corrugated card and its suitability as a packaging material.

Tips: The presentation should be no more than 10 minutes long. Use photos of the products/materials and suitable graphics to enhance the presentation.

Exam questions
AS exam question

1. (a) Name an application for each of the following materials:

    (i)   corrugated card;

    (ii) layout paper;

    (iii) foam board;

    (iv) carton board.   [4 × 2]

(b) Explain why each material is suitable for that application.   [4 × 5]

A2 exam question

1. Paper- and card-based materials are very popular with designers and manufacturers of packaging. For **two** of the following products, describe a specific material that they are made from and explain why they are made from that material. You should make particular reference to the method of manufacture, function, aesthetics and environmental factors:

- milk cartons;
- fast-food packaging;
- cereal boxes.   [2 × 12]

Smart and new materials

Introduction to smart materials

A 'smart material' can be defined as a material whose physical properties change in response to an input. Designers and manufacturers are utilising smart materials in the creation of new consumer products, often making them simpler or even safer to use.

Thermochromic pigments

Thermochromic pigments are colour pigments that can change colour in response to heat. The pigments are typically combined with polymers as plastic products are moulded.

Russell Hobbs' thermocolour kettle

The Russell Hobbs 2001 'Thermocolour' kettle, often referred to as the 'pink kettle', changes from a cool blue colour when cold to a vibrant pink as it boils. In the market for high-tech gadgets, this is clearly desirable and is a strong aesthetic feature used in branding and advertising for this product. The colour change is also quite a good safety feature, as we naturally associate bright colours with danger and heat (e.g. hot-water taps coloured red). Also, if the kettle has been left to stand after boiling the user can instantly see if the kettle is hot enough to make a drink from – thus avoiding re-boiling the kettle.

Baby feeding products, such as bowls, spoons and cups like those pictured made by Tommy Tippee, use thermochromic pigments in their manufacture.

In these examples, the bowls, spoons, etc. are a pink colour when cold but turn bright yellow if the food they are used with is too hot. Prior to using smart materials like these, the consumer would have had to physically test the food/liquid by touching or tasting it. This is inaccurate and unhygienic.

Tommy Tippee feed mug

Thermocolour film

Thermochromic pigments are a special type of liquid crystal which when heated will change colour. The introduction of these pigments in the form of an ink has led to the development of temperature testing strips for medical and other applications, such as testing safe water temperature in baths.

Thermometers

Traditional mercury thermometers, whilst very accurate, are difficult to read and extremely dangerous to use with young children who may bite or break them. Thermochromic pigments can be engineered to change colour across a range of temperatures. So it is possible to make a simple measuring scale that indicates if a patient's temperature is too low, normal or too high, by illuminating the relevant section in a colour. Normally, blue would be used to indicate too cold, green normal and yellow too high.

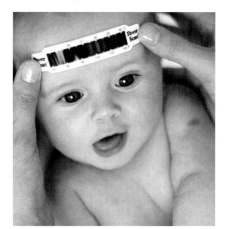

Thermocolour film thermometer

Batteries

Battery test strips incorporate thermochromic ink onto a material that heats up as a current passes. If the battery has sufficient energy to heat the strip, then it will change colour indicating the battery is in good condition.

Battery with test strip

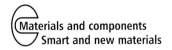

Phosphorescent pigments

Phosphorescent pigments were originally developed as a replacement for the more dangerous radioactive materials and older phosphorescents, used in the manufacture of glow-in-the-dark watch and clock hands.

A new ceramic material has been developed that will absorb large amounts of light energy and will also re-emit this light energy over a period of time. The pigments are available as a powder, and can be mixed with acrylic paints and inks to produce signs that can be seen in complete darkness.

Examples of their use include emergency warning signs, exit signs and watch faces that can be read 24 hours a day.

Shape memory alloys

Shape memory alloys are metals that have been designed to work in a particular way. The most common is called 'nitinol': an alloy of nickel and titanium. Other shape memory alloys can contain a range of other materials such as iron, nickel, cobalt and titanium alloys in varying percentages.

Heat treatment gives the material a memory. For example, a nitinol wire heated by passing a current through it will reduce in length by about 5%. On releasing the current, the wire can be stretched back to its original length. This has applications in bioengineering. For example, a nitinol tube can be collapsed at room temperature and when inserted into a blocked vein will resume its original shape as it reaches body temperature, thereby allowing blood to flow through the vein once again.

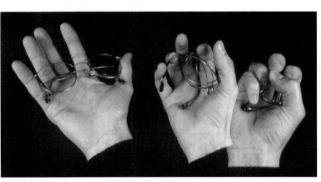

'Memoflex' glasses

A further example, again in bioengineering, includes the plating of broken bones. As the plates reach body temperature, they remember to contract, pulling the fracture together and applying pressure to help bones heal quicker.

'Memoflex' glasses are a further example of the use of shape memory alloys, with their ability to regain their original shape at room temperature after being deformed.

Piezoelectric devices

The piezoelectric effect is achieved by applying an input to a quartz crystal or a polycrystalline material called PZT, causing a desired output to be seen. If a quartz crystal is put under strain, e.g. compressed, a small electrical charge is produced, although the amount of movement in the material is extremely small due to its high stiffness.

Piezoelectric sensors

The charge is proportional to the amount of force being applied, so when conditioned by instrumentation the sensor will give a direct indication of the load being applied. Applications for this type of sensor range from large structures, such as bridges, to sensor mats used in burglar alarm systems.

Piezoelectric actuators

Piezoelectric actuators, on the other hand, work the opposite way round. When a voltage is applied to quartz crystal, a small displacement is produced in the material. The displacement is very small but produces a high mechanical force. A stack actuator – literally a stack of PZT (a common piezoelectric material) – can be used for micro-positioning applications and for fast-acting valves and nozzles.

Other modern materials

Fibre optics

Although considered a new technology, fibre optics was first discovered as long ago as 1870. The first commercial application was around 1966, transmitting information using glass fibres. A good deal of development into reducing the losses found in fibre-optic cables has been carried out since then, enabling data to be transmitted over long distances.

Fibre-optic cables work by transmitting light as a series of pulses, i.e. '0's and '1's. An example of how light travels through glass fibres can often be seen in decorative lamps.

Significantly greater quantities of data can be transmitted than when using copper wires. Each glass fibre in a fibre optic can carry data and, when multiplexing is employed, one fibre alone could replace hundreds of copper cables.

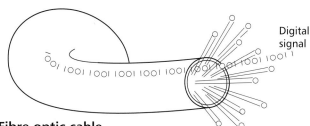

Copper cable
Could be a single cable or multi-stranded, yet only one signal

Fibre-optic cable
Numerous strands of fibre optics enabling numerous signals to be sent

A transmitted television signal is converted from an electronic signal to a light signal by the transmitting modulator, which changes the signal from its analogue source to a digital format where the signal is either on or off ('1's or '0's). At the receiving end of the fibre-optic cable – a digital TV for example – the signal is once again converted to electronic data.

Liquid crystal displays

Liquid crystal displays (LCDs) have been around for a number of years; they can be found in a variety of electronic products, including:

- the alpha-numeric displays in calculators and electronic dictionaries;
- mobile telephones;
- PDAs and laptop computers.

Liquid crystals are carbon-based compounds, which when aligned in their 'natural' form will allow light to pass through. If, however, a small voltage is applied to the crystals, then their orientation is changed to match the path of the electron flow so blocking light. This is what makes them appear black on the display.

LCDs can be reflective or backlit. Reflective displays reflect the light entering the display back out again. Calculators generally have reflective displays. Mobile telephones and laptop computer displays are usually backlit, making them usable in poor light conditions.

Optical fibres

Genetic modification of woods

Research and experimentation into genetically modified timbers has been growing rapidly. The aim of this research is to provide:

- quicker growing trees;
- trees that are more resistant to rot and insect attack;
- trees with reduced lignin.

The benefit of producing quicker growing trees is in aiding the efficient management of forests by replacing trees as they are being cut down.

Trees more resistant to rot and decay and insect attack increase the life of timber products. There is a drawback with this, in that traditionally some timbers may have been left on the forest floor to rot. The new timbers that are rot-resistant may have to be burned, resulting in further emissions that will have an effect on the ozone layer.

The use of trees with reduced lignin will be beneficial in reducing the necessity for removing this tough natural resin with the toxic chemicals currently employed. Another benefit of having trees with reduced lignin is that the wood can have various degrees of colouring, which could be employed in the design of furniture and external products.

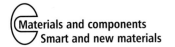
Biodegradable plastics

Conventional oil-based plastics do not break down easily and, since the main bulk of domestic waste is made up of plastics, they have a significant effect on the environment. Biodegradable plastics however are designed to be degradable under the right conditions, i.e. in a biologically rich environment.

Biodegradable plastics are derived entirely from renewable raw materials. For example, starch from wheat, corn and potatoes is a natural polymer that can be modified to form a plastic called polylactide (PLA).

PLA can be used in the manufacture of plant pots and disposable nappies. Certain types of PLA have been used successfully in medical implants and in sutures, because of their ability to dissolve over time.

Biodegradable plastics can also be produced from bacteria. These polymers are called polyhydroxyalkanoate (or PHA). The plastic is harvested from bacteria growing in cultures.

Both these methods of producing plastics are currently expensive, resulting in products manufactured from them being up to 10 times more expensive than products from traditional oil-based plastics.

PRODUCT ANALYSIS EXERCISE: *smart and new materials*

Tefal Red Spot pans

1. The Tefal Red Spot range of pans uses a smart material to make a red dot in the centre of the pan. Explain what type of smart material is used and how it functions.

2. Explain what function the red dot has.

3. Describe how similar smart materials are used in other products that you are familiar with.

Exam questions

AS exam question

1. (a) Define what is meant by the term 'smart material'. [3]

   (b) Name a specific smart material. [2]

   (c) Name a suitable application for that material and explain why it is suitable. [8]

   (d) Use notes and diagrams to show how the product is manufactured. [10]

   (e) Describe the health and safety measures that you would take when making the product described in part (d). [5]

A2 exam question

1. Designers and manufacturers are making increased use of smart materials to enhance the performance of products. For **two** of the following product areas, describe specific examples of where smart materials are used to enhance the performance of the product:

   • Kitchen products, e.g. kettles, saucepans, etc.

   • Children's toys.

   • Medical equipment. [2 × 12]

Processes and manufacture

Introduction to Part Two

As a student of Product Design, you need to know about how materials and components are processed to create the product from the design. Many of the processes you need to know about have already been covered in Part One, where they related to particular materials and components. Part Two considers some further processes relating to joining, the deterioration of materials, finishes and finishing processes, and testing materials.

You also need to know about the general nature of manufacturing systems used to produce commercial products; the second half of Part Two provides key information for you on manufacturing systems, safety in manufacturing, and the application of IT in industrial and commercial practice.

As in Part One, you will need to use other resources – books, CD-ROMs, the Internet, etc. – to give you all the information you need for success as a product designer. Some references for further reading are provided here.

Joining processes

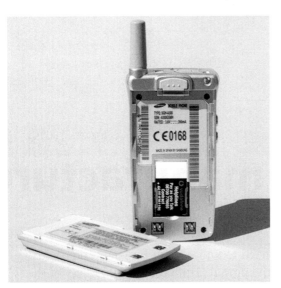

Moulded fittings in mobile telephones

Introduction

We will be taking a look at appropriate joining processes available for use with a range of materials.

Products such as mobile telephones have components that are manufactured using a variety of processes from a range of materials. All of these components need to be assembled together to create a single functioning product. To achieve this, a variety of joining processes will have been used.

For example, removing the battery cover of this mobile telephone simply requires it to be pushed out of its locations to release it. The photograph also shows that the battery cover can be replaced by reversing the above. This is achieved by integrating the joining method into the two components to be joined.

Other joining processes require the inclusion of an extra component or material. Examples include nuts and bolts used to clamp two or more components together or the use of filler rods when welding components together.

At AS level

As an AS level student you should be aware of the general types of joining process and whether the processes are temporary or permanent. You should begin to know and understand the differences between soldering, brazing and welding, and how they are carried out. You should also understand the basic types of joint for wood, including knock-down joints, and how they might be used. Knowledge of joints that are created with the use of screw threads in their many forms (nuts, bolts, self-tapping screws, captive nuts) is also desirable, as is a range of adhesives and appropriate applications.

At A2 level

A2 students should build on the knowledge gained at AS level to give reasons why a particular joining method has been chosen, the advantages and disadvantages of that method, along with an understanding of how the choice of material clearly affects the type of joining process used.

A2 level students should also be aware of the effect a joining process may have on a material and possible alternative methods to counter this.

Choosing a joining method

Before we can say exactly which joining method we are going to use, there are a number of important factors that must be considered. The most important, of course, is the material – a wood cannot be welded, for example.

- **Temporary or permanent:** will the joint need to be dismantled at all?
- **Joint strength required:** how strong does the joint have to be? Will it be under high levels of stress?

- **Stiffness needed:** does the joint need to be flexible in any way, or should it be rigid as in a frame structure for a piece of furniture?
- **Effect of joining method on materials being joined:** will materials being joined be adversely affected by, for example, heat from welding, soldering or brazing?
- **Appearance around the joint:** will the joining process affect the surface finish? Will extra processes be required to 'clean' the joint?

Once these factors have been considered, then an appropriate choice of joining method can be made.

Temporary or permanent joining processes

Joining methods fall into two broad categories – those that are temporary and those that are permanent.

- Temporary joining methods are those that do not damage the materials being joined when the joint is undone. The most common example of this is the nut and bolt arrangement.
- Permanent joining methods are those where dismantling them will damage the materials being joined. A good example of this type of joint is the spot-welded joints found on car body panels, which are intended to be extremely rigid and last for the lifetime of the car.

Summary of joining processes

The diagram below highlights a good range of joining processes.

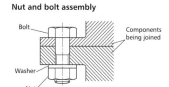
Nut and bolt assembly

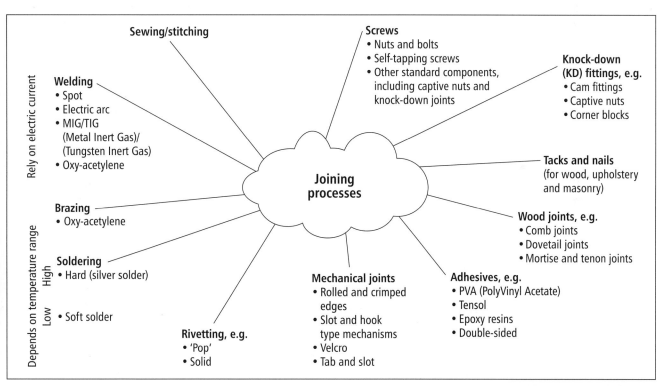

Temporary joining methods

We have already seen that nuts and bolts are a common method of joining components together. However, there is a large range of different types of standard component that carry out a similar function. These include wood screws and self-tapping screws, which cut their own threads as they are screwed into a material. If a metal, like steel, is thick enough, a thread can be formed in an appropriately sized hole removing the need for a nut.

For sheet material, specially shaped spring steel nuts are used with self-tapping screws, whereas with woods a range of different nuts made from

Captive nut in pine frame

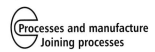

plastics or metals for use with bolts and screws are available. Collectively, these are called 'captive nuts', because they are held in position in or on the material.

Knock-down joints make extensive use of captive nuts. These products are so called because they were originally designed for use with flat-pack furniture. A range of knock-down fittings are shown in the photograph. Notice that they make use of a range of threads and cams depending on their intended use.

Joining metals

Metals can be joined in a number of different ways. We have already discussed nuts, bolts and screws. Other methods, like folding and crimping, involve the forming of the metal to provide a joint; while others, such as soldering, brazing and welding, require the application of heat and, in some cases, the inclusion of a filler rod.

Processes involving forming

Some very simple joints can be made in sheet metals by folding the materials together. An example of this can be found in the sweet tin shown. The seam down the length of the container is produced by 'hooking' the two edges together.

The base of the container is secured by crimping the folded edge on the base to the bottom edge of the container. This is the same method used to secure the aluminium top of a soft drinks can to the main body.

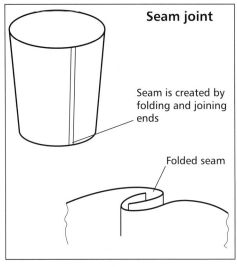

Seam joint

Seam is created by folding and joining ends

Folded seam

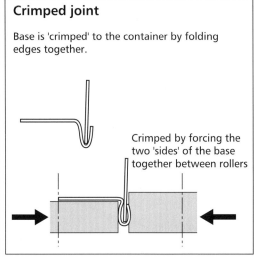

Crimped joint

Base is 'crimped' to the container by folding edges together.

Crimped by forcing the two 'sides' of the base together between rollers

Processes involving heat

The range of processes covered by this section includes:

- soldering;
- brazing;
- welding.

They all involve the fusion of the materials being joint or of a filler material.

Soldering

- Soldering has the lowest temperature of these joining processes.
- There are a range of solders available from soft solders to hard solders.
- Soft solders are used for joining electronic components to circuit boards, or for small copper or brass components.
- Hard solders are used for products requiring additional strength and to aid the construction of products requiring the soldering of a number of mating components.

- Soft solder melts at around 200 °C.
- Hard (silver) solders are so called because they contain a small amount of silver that is alloyed with copper and zinc. The amount of each material differs to produce a range of hard solders that melt at temperatures between 600 °C and 800 °C.
- Traditionally, solders have contained lead in the ratio 60% tin and 40% lead. Nowadays solders are available that are lead-free, containing 96% tin, 3.5% silver and 0.5% copper; consequently, these are more expensive than traditional solders.

Making a soldered joint

Whichever type of solder is to be used to make a joint, the process remains basically the same.

Stages of the process:

Step 1 Materials to be joined must be cleaned to removed grease and dust. The surfaces to be joined should be kept clean with the use of a flux.

Step 2 The mating surfaces must fit together without there being large gaps, but must be held together securely while being heated up.

Step 3 A blowtorch is used to heat the material around the joint, ensuring both pieces are heated evenly. The solder filler rod is rested on the joint – the heated material will melt the solder and capillary action will allow the solder to run between the joint.

Step 4 As soon as the joint is completely filled, it should be allowed to cool.

Soldering

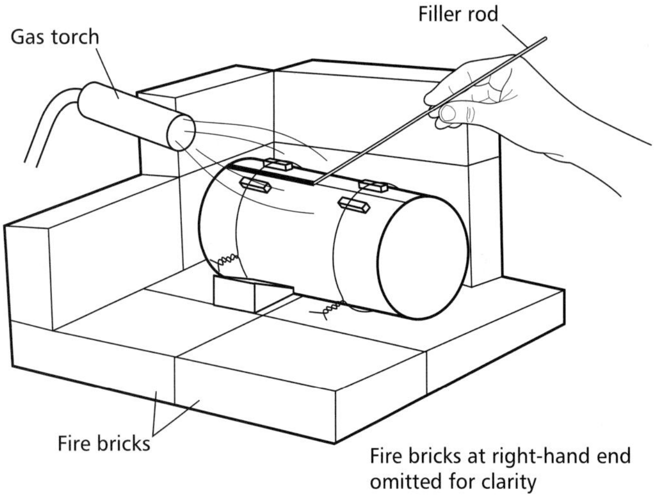

Filler rod

Gas torch

Fire bricks

Fire bricks at right-hand end omitted for clarity

Brazing

- Brazing takes place at a higher temperature than soldering.
- The filler rod in this case is a brass alloy, called brazing 'spelter' and melts around 880 °C.
- The materials that can be joined using this process include copper and steel (in particular mild steel).
- The process is essentially the same as soldering, with the materials being joined cleaned and kept clean using a flux – in this case Borax is used. Components are held together while being heated.
- When the correct temperature is reached, the brazing rod melts at the joint filling the joint by capillary action. Again the material is allowed to cool before having excess flux and braze removed.

Welding

Welding differs from soldering and brazing, in that the materials being joined must be the same and if a filler rod is being used then it is the same material as that being joined.

Methods of welding

There are a number of different methods of welding, including the use of oxy-acetylene or an electric arc to generate the heat required. Other techniques use an electric current passing through the materials to fuse them together.

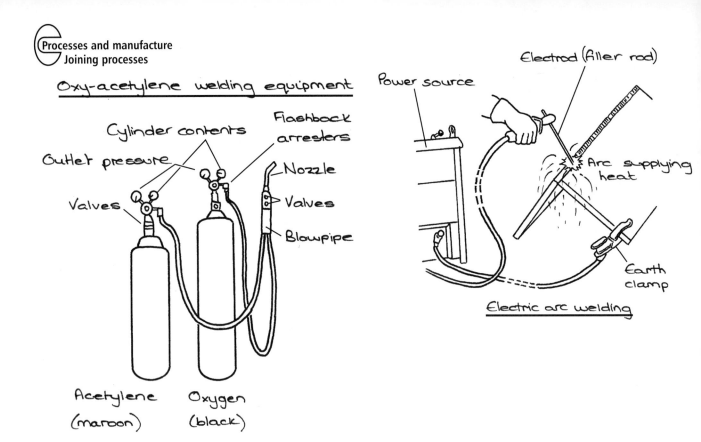

Oxy-acetylene welding equipment

Cylinder contents

Outlet pressure

Flashback arresters

Nozzle

Valves

Valves

Blowpipe

Acetylene (maroon)   Oxygen (black)

Power source

Electrod (filler rod)

Arc supplying heat

Earth clamp

Electric arc welding

- **Oxy-acetylene welding:** this uses a mixture of oxygen and acetylene to create a flame that will burn at a temperature of around 2500 °C at the hottest point. This will clearly be sufficient to melt mild steel at the joint, allowing the melting of a filler rod to fuse the joint edges together.

- **Electric arc welding:** this generates sufficient heat to melt the joint edges by creating an electric current through a gap (arc) between the materials being joined and the filler rod (electrode). The electrode is coated in a flux which, when melted, prevents the joint area becoming oxidised.

- **MIG and TIG welding:** more refined forms of electric arc welding can be used to join thin sheet material. MIG (Metal Inert Gas) and TIG (Tungsten Inert Gas) welding use a gas jet around the filler wire to prevent oxidation of the material. Different gases are used with different materials, for example argon is used with aluminium.

- **Spot and seam welding:** both spot welding and seam welding use an electric current as the heat source. Spot welding, as the name suggests, provides a spot of heat to fuse the metals together. It is usual to find a series of spot welds in a structure like a car body shell. Seam welding, on the other hand, passes an electric current through the material as it passes under rollers. A typical application for this type of joining method is tin-coated mild steel for food and drinks cans.

Using spot welding to join sheet metals

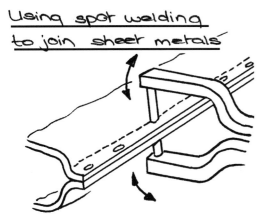

Spot welding on a car body shell

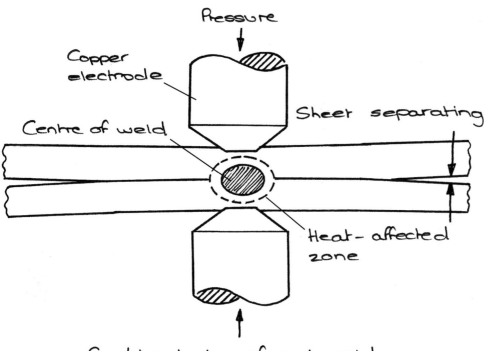

Pressure

Copper electrode

Centre of weld

Sheet separating

Heat-affected zone

Sectional view of spot weld

Seam welding

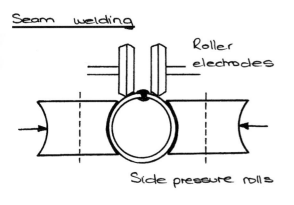

Roller electrodes

Side pressure rolls

Example of a seam welded product

Task

Identify the safety precautions that must be observed when joining metal components by (a) oxy-acetylene welding and (b) electric arc welding.

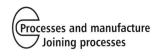
Joining processes for woods

A variety of both temporary and permanent joining methods can be used with woods, depending on the intended application.

- When large sections of timber are used – for example, the oak framework for a house – the beams and posts can be joined together by a combination of traditional wood joints and pegs.

Finger joint

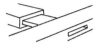
Mortise and tenon joint

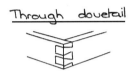
Through dovetail

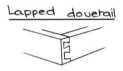
Lapped dovetail

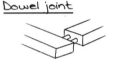
Dowel joint

Wood joining methods

- When large section beams are used for temporary support, however, a more temporary joining arrangement may be used. In the case shown above, a barn wall is being supported by a framework that is bolted together.

- A modern roof truss is produced in a different way, i.e. without the use of traditional joints. Each of the timbers is cut to the correct size and shape and, once set up in a jig, plates with spikes (effectively nails) are forced into both sides of the joints.

- For traditional household furniture, more traditional wood joints would be used. The type of wood joint again depends on the application.

Note All of these joints would be secured with an adhesive, such as PVA (Poly Vinyl Acetate), making them permanent.

Self-assembly furniture

Self-assembly furniture is produced in high volume, being supplied to the consumer as a flat-pack of component parts. In order to accommodate self-assembly, a range of fixings has been developed specifically for use with this

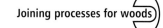

type of furniture. With every flat-pack there is a pack of knock-down (KD) fittings included. The joining methods used may comprise any or all of the following: captive nuts, cam fittings and corner blocks, depending on the materials used in the manufacture of the furniture, i.e. kitchen units are generally made from laminated chipboard, while bedroom furniture may be made from solid pine. Examples of these joining methods are shown below.

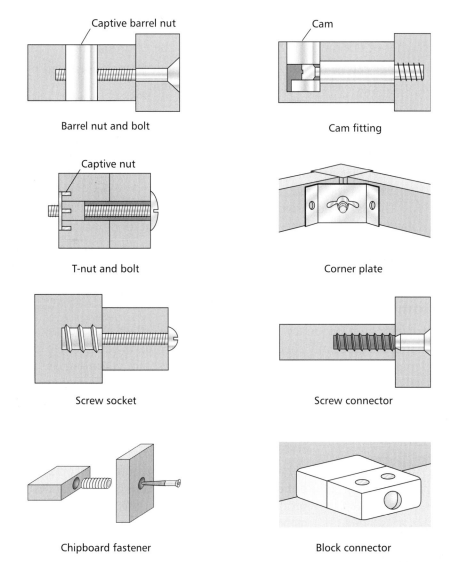

Barrel nut and bolt

Cam fitting

T-nut and bolt

Corner plate

Screw socket

Screw connector

Chipboard fastener

Block connector

A range of knock-down fittings

Along with these fixings there will be:
- hinges that have been especially made to fit into the pre-drilled holes;
- a range of screws specially developed for use with self-assembly furniture: different screws will be used for chipboard products from the standard screws used with natural woods.

All of the joining components, including hinges and handles, are manufactured in large numbers by specialist companies and, although originally designed for the specialist application of self-assembly furniture, are very common. These are standard components and can be readily purchased from local suppliers.

Wood adhesives

Where a joint is to be permanent – whether it be a traditional wood joint in natural timber or wood products assembled with knock-down fittings – an adhesive can be used to aid the integrity of the joint.

The adhesive used would generally be PVA. This would be spread on the two halves of the joint that have already been cleaned, and the whole joint clamped together until the adhesive is dry. Any excess can be wiped away with a damp cloth before it has chance to dry.

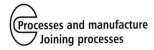
Joining polymers

Mechanical means

Polymers can be joined together by mechanical means, i.e. nuts and bolts. The choice of material from which nuts/bolts are made depends on:

- the application;
- the type of polymer to be joined;
- the amount of load the joint is to carry;
- whether or not the joint will be affected by the environment.

Standard fixings can be obtained in mild steel, stainless steel, brass and nylon; used appropriately, these will produce a durable joint.

Pre-cooked food packaging

Adhesives

Adhesives can be used to join polymers but, in general, are not wholly successful. Tensol is an adhesive in liquid form that is used especially with acrylics. It acts as a solvent, by dissolving the components together at the joint.

The use of adhesives in packaging is more successful. When heat is applied directly to the film lid of a pre-cooked food package, a layer of adhesive is melted on the rim of the container allowing the film lid to adhere to the container. This also creates a seal between the two components making the packaging hygienic.

Ultrasonic welding

The use of very high frequency sound waves is an excellent method of joining plastic materials, especially sheet materials. The two parts being joined are firmly clamped together. Very high frequency (ultrasonic) vibrations are then introduced to the materials through the clamp. This has the effect of generating heat, which is produced by the vibration of the atoms and molecules of the materials being joined.

Integral fixings in a CD case

Integral fixings

Polymers can be used to manufacture complex three-dimensional shapes needed to produce modern products. The advantage of this is that fixings, such as posts for screws, captive nuts, locating and securing clips, can all be made integral to the component being joined.

Examples include the battery cover, 'express-on' covers for mobile telephones, and the clips that hold a CD into its case. These types of fixings rely on the elastic properties, as well as the strength and durability, of the polymers used to make the components.

Joining ceramics

Ceramic products, such as teapots and cups, are manufactured from the same material throughout – whether it is an earthenware body or a fine bone china. The handles on these products will have been stuck on while the clay material was in the 'green' state, i.e. before firing. The 'adhesive' used in this instance would be a liquid clay (called slip) – a much watered-down version of the clay used for the product. Once the handle is fixed on the body of the teapot or cup, then the whole thing is fired to produce a single piece.

Adhesives

The main advantage of using adhesives as a joining technique over any other method of joining, is that it is essentially invisible. Adhesives do not damage or change the shape of the components being joined, whereas nuts and bolts are visible and spot welds can cause indents in the metals being joined, for example.

There are two main groups of adhesives:

- **natural adhesives** include any of the animal- and vegetable-based glues, and naturally occurring resins like gum arabic;
- **synthetic resins** include manufactured adhesives, such as epoxy resins, phenolic and formaldehyde resins, anaerobic adhesives and the silicon-based adhesives.

Table 17: Examples of adhesives and where they might be used

	Type	Examples	Uses
Natural adhesives	Animal glues	Animal hide, bones, hooves	Used to glue woods, fabrics and leathers
	Natural resins	Gum arabic	Used for papers and fabrics, and binders in watercolour paints
	Inorganic cements	Portland cement	Used in the building industry for bonding bricks and blocks
Synthetic adhesives	Synthetic resins	Cascamite (powder mixed with water)	For bonding woods; is waterproof and can fill slight gaps
	Epoxy resins	Araldite Two (part adhesive, hardener and resin)	Joins most materials
	PVA	Poly Vinyl Acetate (water-based)	Used for gluing woods; generally not waterproof
	Contact adhesives	Evo-Stik (works by evaporation of solvent)	Used for bonding sheet materials on contact
	Hot glue sticks	(Work on application of heat)	For rapid bonding of papers and cards
	Acrylic cement	Tensol	For gluing acrylics only

Using adhesives to join materials together requires good surface preparation, i.e. surfaces thoroughly cleaned of dust, grease and any previous coatings. The adhesive should then be applied to the components, either one or both – depending on the type of adhesive being used. The components being joined will then have to be held together until the adhesive has dried.

Advantages of using adhesives

- They are able to join dissimilar materials.
- The insulating properties of adhesives help prevent corrosion through electrochemical action of dissimilar metals while some adhesives, particularly the silicon-based varieties, act as sealants as well as bonding agents. Examples of this can be seen on vehicle windscreens, where the old rubber seal has been replaced by a silicon-based adhesive which also acts as a sealant around the windscreen.

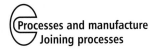
Task

1. Identify the safety precautions that must be observed when using adhesives. Consider at least three different types of adhesives, e.g. PVA, epoxy resin, contact adhesive.

2. Compile a table of common adhesives, including brand names and common applications that the adhesives are used for.

3. Identify two different adhesives that are hazardous. Explain what health and safety precautions you should take when using them.

Case study: Joining processes

A range of temporary joints

The Dyson DC01 vacuum cleaner

The DC0I vacuum cleaner is a good product to study, as it shows a range of temporary joints that are used throughout the casing and the functional parts of the cleaner.

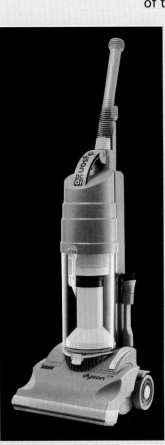

Dyson DC01 vacuum cleaner

- If the underside of the vacuum is examined, we can see that the cover that fits over the brush is fixed using self-tapping screws. These screws enable the cover to be removed so that blockages can be cleared and, in the event of the drive belt breaking, it can be replaced easily. As the screws have standard Philips heads, they can be removed with a common screwdriver. No specialist tools are needed.

- The DC01 features a number of parts that use 'click' fittings. Two parts in particular are the handle and the dust canister. The handle on the DC01 is removed from the vacuum as it forms a 'wand' that can have tools inserted into it to enable vacuuming of stairs and other inaccessible parts. The handle features a squeezable part that disengages a click fastening from the main body of the vacuum. The dust canister uses a spring-loaded fastening to attach the canister to the main body of the vacuum.

- Within the motor housing, and other sections that contain electrical parts, the manufacturer may use temporary fastenings such as machine screws but they may have torx heads which require a torx driver to undo them. This helps to prevent unqualified consumers tampering with parts of the cleaner that could lead to electrical shock. However, temporary fixings are still required to allow for maintenance of such parts.

- The casing of the DC01 is made from recyclable plastics. Dyson has a facility that enables owners to return their cleaner at the end of its useful life, so that the plastic parts can be shredded and later recycled. The use of temporary fastenings such as self-tapping screws and machine screws enables all of the parts to be disassembled relatively easily.

- Very few, if any, permanent joints are used in the manufacture of the DC01. The casing is injection moulded and therefore fewer joints are necessary. Basically, the more joining that has to be done in the manufacture, the longer it takes to make, increasing cost. Where it is necessary to join plastic parts together, they are normally joined with temporary joints such as those described above.

PRODUCT ANALYSIS EXERCISE: *joining processes*

Aluminium bicycle frame

Study the photograph of an aluminium bicycle.

1. The aluminium bicycle frame has been fabricated together. Use notes and diagrams to explain how this is done.

2. If the bicycle frame had been made from mild steel tube, what joining method would have been used? Explain why the same joining method can't be used.

3. The bicycle seat is adjusted using a temporary joint. Use notes and diagrams to show how this temporary joint functions.

PRODUCT ANALYSIS EXERCISE: *joining processes*

Stainless steel saucepan

The saucepan shown is made from stainless steel.

1. Explain why stainless steel is a suitable material for manufacturing a saucepan.

2. Name the process for manufacturing the body of the pan.

3. The handle has been spot welded on. Explain why this process is appropriate.

4. Use notes and diagrams to explain how the spot-welding process works.

Exam questions
AS exam question

1. (a) Use notes and diagrams to show how two different KD fittings work. [2 × 5]

   (b) Explain the benefits of using KD fittings for :
     (i) the manufacturer;
     (ii) the consumer. [2 × 5]

   (c) With reference to your own coursework or another product you are familiar with, describe how the product was made using fabrication techniques. You should use diagrams to support your answer. [8]

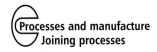

A2 exam questions

1. Products can be manufactured by either redistribution, wasting or fabrication processes. Discuss the merits and disadvantages of each type of process, making reference to specific products that you are familiar with. [24]

2. Explain the joining methods used to manufacture **two** of the following products:
   - car body panels;
   - domestic appliances such as washing machines;
   - flat-pack furniture.

   You should describe the type of joining method used and the reasons why each method is used. [2 × 12]

Corrosion, decay and degradation

Over time and under varying conditions materials will eventually begin to break down. Here we will concentrate on what happens to metals, woods and plastics.

At AS level

As an AS level student you need to become aware of these processes and use the correct terms when discussing them in your answers to examination questions. For example, metals are said to 'corrode', woods 'decay' and plastics 'degrade'. This will be the approach of this unit, with descriptions of how these processes take place and what happens to the material in question.

At A2 level

A2 students should build on their understanding of detrimental effects of the environment in which various materials or combinations of materials are found, and the mechanisms that take place as a material deteriorates.

Corrosion of metals

Most metals have an oxide layer on the surface of the material. This is always present and is the result of the way oxygen in the atmosphere reacts with the material at the surface. Generally, this oxide layer is helpful to the material by being sufficiently dense and impermeable to prevent further oxidation. If we take a non-corroding material, such as brass, and purposely scratch the surface exposing 'new' metal the oxide layer closes up immediately and again helps protect the material.

Rust

Ferrous metals, with the exception of stainless steel, are the only group of materials that rust. This type of corrosion is caused by the material's oxide layer being porous, allowing moisture to make contact with the surface of the material. This in turn causes the surface of the material to rust, i.e. to become the familiar red-brown colour. Moisture will penetrate the initial layer of rust, which, in turn, will oxidise a further layer lifting up the previous layer.

Task

Clean a piece of mild steel sheet so that it shines. Hold your thumb or finger onto the cleaned area for a few moments. Leave it on a shelf for 24 hours. When you come back to it, what do you see? Probably a rusty brown fingerprint. How has this happened?

Electrochemical corrosion

In this type of corrosion, a very small electrical current is produced when two different metals are joined together. All metals have a natural voltage. When two different metals are joined together in the presence of rainwater (and the acids it might contain), then an electrochemical cell is produced and one of the materials will begin to corrode. The diagram helps to explain this.

Electrochemical corrosion can be very slow because of the very small voltages and currents involved, but over time corrosion will take place.

How a battery works

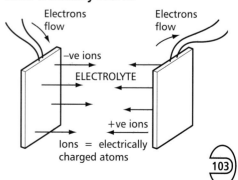

Processes and manufacture
Corrosion, decay and
degradation

For example, if brass screws were used to hold aluminium sheets together, there will be the basis for an electrochemical cell. If rainwater comes into contact with this join, then an electronic circuit is produced because of the voltage difference between the two materials. Aluminium is the material that will begin to break down at the junction of the two materials. The brass itself will remain generally unharmed, though it may be difficult to separate from the aluminium.

> ## Task
>
> Review the unit on joining processes. What method of joining aluminium sheets would be more beneficial than the use of screws? Explain the reason for your choice.

Corrosion through chemical attack

Acids have a corrosive effect on most metals and can rapidly destroy a metal component. This is often associated with extremely dangerous forms such as hydrochloric acid. These types of liquids are generally held in plastic containers, since polymers have a generally high resistance to them.

> ## Task
>
> Surprisingly enough, soft drinks may contain some potent acids – though these are not generally harmful to the consumer. Pour cola, or a similar drink, into a clear plastic container followed by a few old copper coins. Leave for 24 hours or more before removing them from the liquid. What has happened to them? Identify the chemicals in the liquid that might be responsible for the change.

This will, of course, have repercussions for the material used for the container. Glass and plastics, like PET (polyethylene terephthalate), are resistant to acids. However, aluminium cans have to be protected so that the liquid does not corrode the container and that the container material does not contaminate the drink. This is achieved through coating the aluminium with a thin layer of polymer.

Decay of woods

Being a natural material, woods are open to attack from a number of fronts:

- wet rot;
- dry rot;
- attack by insects.

Wet rot

Wet rot is where timber endures alternating wetness and dryness and begins to decompose. Moisture is absorbed into the timber where it will partially dry out, followed by more moisture absorption resulting in the resins and fibres breaking down. A protective barrier must be applied to prevent the ingress of moisture.

Dry rot

Dry rot is a fungus that spreads its strands very quickly through woods, e.g. in a building. It is called 'dry rot' because of the way it converts the timber into a dry, soft, powdery state. The fungus thrives where the conditions are damp and unventilated, with little circulating air. Increased ventilation will help prevent dry rot in the first instance, but where it has already attacked timbers they should be replaced.

Insect attack

The furniture beetle (or woodworm as it is more commonly known) can be responsible for attacking softwoods and hardwoods in floorboards and furniture. The deathwatch beetle will generally only attack hardwoods, e.g. the structural timbers of old buildings.

Insects lay their eggs in a crack or crevice in the timber. The larvae that hatch eat their way into the fibrous structure of the wood creating tunnels. When the time comes for the grub to pupate, it returns to a cavity that is near to the surface. When the beetle finally forms, it eats its way out through a flight hole.

Moderate attacks by woodworm or deathwatch beetle can be treated with chemicals, but where the damage is severe, the affected timbers should be removed and burned to prevent further contamination.

Degradation of plastics

Many plastics degenerate in the environment. To some degree this is simple oxidation, but is more seriously caused by ultraviolet radiation.

Stability of plastics

Many plastics are relatively inert and will resist chemical attack. For example, although polythene is unaffected by prolonged contact with concentrated acids, including hydrofluoric acid, it is less stable when exposed to outdoor environments. In this case it will tend to become more brittle and opaque.

Some rubbers used in hoses and seals can perish – becoming brittle and useless if not stabilised.

Weathering of plastic materials

Unless stabilised, almost all polymers, and in particular thermoplastics, will deteriorate in appearance. This seems to be due mainly to the combined effects of oxygen and ultraviolet light.

Most polymers absorb UV light causing the chemical bonds in the molecular chains to break and so shortening them. At the same time, the oxygen in the atmosphere leads to the formation of chemical groups and possible cross-links. Both of these events cause the material to become less flexible and weaker.

A stabilising substance that will shield the material from UV radiation, by making it opaque, is added during manufacture. Pigments such as 'carbon black' – used in the tyre industry – will absorb UV radiation and thereby act as an effective screen.

Example of weathering

Take, for example, the effect of weathering in the material being used for glazing a greenhouse. The 'glazing' can be made from thin sheets of LDPE (low density polyethylene). This will last for a couple of seasons, before it will start to degrade due to exposure to very strong UV light from the Sun. The overall cost of replacing the LDPE is lower than using glass as the glazing medium, which would also require a stronger frame to take the heavier material.

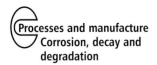
Processes and manufacture
Corrosion, decay and
degradation

PRODUCT ANALYSIS EXERCISE: *corrosion, decay and degradation*

Iron/wooden gates; plastic playhouse

1. Study the photograph of the wrought iron gates.

(a) Explain why the wrought iron would corrode if the gates were not given a protective finish.

(b) Describe a suitable protective finish and explain in detail how it would be applied.

2. Study the photograph of the wooden gate.

(a) Explain why wood generally decays if it is not given a protective finish.

(b) Describe a method that could be used to prevent the wood from decaying.

3. Study the photograph of the plastic playhouse.

(a) Describe how some plastics degrade.

(b) What methods would the manufacturer use to ensure that the plastic used in the playhouse does not degrade?

Exam questions

AS exam questions

1. Describe how ferrous metals corrode. [4]

2. Name a specific finish that can be applied to ferrous metals to prevent corrosion. [2]

3. Explain in detail how the finish is applied. [6]

4. Name a specific finish that is applied to metals for aesthetic purposes. [2]

5. Describe in detail how that finish is applied. [6]

6. Name a specific polymer that is resistant to weathering processes. [2]

7. Explain what makes the polymer resistant to weathering. [6]

A2 exam question

1. (a) With reference to two different products made from wood that you are familiar with, explain how specific finishes are applied to prevent decay. [12]

(b) With reference to two different products made from metals that you are familiar with, explain how specific finishes are applied to prevent corrosion. [12]

Finishes and finishing processes

Introduction

At AS level

As an AS level student, you need to be aware of the range of finishes that can be applied to products and the reasons why these processes are necessary. The difference between finishes and finishing processes must be defined, along with an understanding of terms such as self-finishing and self-coloured.

At A2 level

A2 students should again build on the knowledge and understanding gained at AS level, being prepared to provide full explanations of finishing processes and their appropriateness for the application.

Why include a finishing process?

The finish of a product significantly affects its aesthetic qualities, giving it a greater sense of quality and value and therefore adding to its appeal.

There are a variety of reasons, as well as aesthetics, why a finish should be applied to a product including:

- protecting the material against corrosion;
- making the product water repellent;
- improving its resistance to wear and fatigue;
- improving its ability to reflect or absorb heat;
- improving its ability to insulate against heat or electricity.

Finishing materials

Self-coloured

One of the advantages of using plastics to manufacture products is that it can be self-coloured. By including pigments with the powders or granules in the hopper of the moulding equipment, a product component, e.g. casing for a CD or mobile telephone, can be produced in the desired colour.

Textures

Textures can be added to polymers by applying the texture to the surface of the mould. This is directly transferable to the polymer because of the way the softened material is forced under pressure into every part of the mould. Conversely for a smooth finish, the surface of the mould would have a correspondingly smooth finish. These processes will clearly add expense in the production of the mould.

Self-finishing

Plastics are also known as self-finishing materials. This means that the quality of the mould must be very high to ensure that any trimming is kept to a minimum.

Cutting processes

Minimum trimming is not always the case with metals and woods where, following cutting processes, some burrs (rough edges) must be removed. This means an additional process and ultimately additional manufacturing cost.

Modern industrial metal blanking and piercing processes used to manufacture, for example, vehicle seat components from flat sheet, have been developed to reduce the need for deburring.

Further developments in metal cutting have been made in recent years in the use of laser and plasma cutting. These cutting processes are controlled by computer and are generally used for accurately cutting thick sheets of metals. The photographs on page 43 shows the quality of finish available from these processes. The two samples are both 12 mm thick toughened steel and have been cut from components destined for a JCB digger.

Finishing processes for metals

Natural barriers

Metal oxides already exist on most metals and, in the majority of cases, are a barrier against environmental effects. This is not the case with steels, where the oxide layer is porous and allows moisture to come into contact with the metal – so it continues to oxidise (rust). Stainless steel is the only type of steel that is non-corrosive. This material has chromium (as well as other elements) added, providing a protective layer of chromium oxides.

A useful oxide layer can be formed on steels by 'blueing' them. This involves heating the steel up to 300 °C and then quenching in oil. The result is a fine oxide layer that helps protect the material against atmospheric effects.

Applied barriers

As we have seen, steels in particular need protecting from corrosion. There is a large range of materials that can be applied to metals, by either:

- brushing;
- rolling;
- spraying;
- dipping.

Electroplating

Other metals can be used to coat the base metal. The method used is electroplating, an electrochemical process that allows ions from the coating material to form on the base material, giving it the finish of the coating metal. An example of this is chromium plating that is used to enhance the properties of bathroom taps, kitchen equipment and vehicle components.

Metals that can be used as coating materials in this way include:

- gold;
- silver;
- tin;
- zinc;
- copper.

Anodising

Natural barriers can be enhanced. Aluminium, for example, can be anodised. This process makes the surface of the aluminium more durable and resistant to scratches. Anodising is produced in an electrochemical cell where a sulphuric acid solution is the electrolyte, with the aluminium product as the anode and lead as the cathode. Passing an electric current through the electrolyte solution effectively builds up a tough oxide layer that will accept dies. It is finally sealed with a lacquer.

Processes and manufacture
Finishes and finishing
processes

Dipping and spraying

Metals can be coated with other metals by dipping or spraying in conjunction with some kind of heating system. The materials need to be cleaned, then fluxed prior to applying the coating material. Examples include:

- **Zinc plating:** a galvanising process achieved by dipping steel in molten zinc heated to 450–60 °C, and is usually the first layer of protection for car bodies;
- **Tin plating:** tin is applied by passing the sheets of steel through baths of molten tin at 315–20 °C. This is used in the manufacture of food cans.

Enamelling

Vitreous enamelling is a process where finely ground glass, formed into a water-based slurry, is sprayed onto a metal product, e.g. components for cookers or for pressed steel baths. It is then fired, so that the coating becomes a continuous layer of heat- and scratch-resistant material.

Painting

For a paint finish to be applied successfully, the metal surface must be thoroughly cleaned and degreased using a paraffin-based liquid. This ensures that the primer (the first layer of paint) keys into the material's surface. The second layer will be an undercoat. This helps give a professional finish to the topcoat. Paints can be brushed on or sprayed.

Car bodies are generally made from mild steel. Nowadays they last a lot longer before corrosion takes over, primarily because of all the layers of protection that have been applied to them. Following the galvanising layer (described above) numerous layers of primer, undercoat and topcoats of acrylic or cellulose paints are sprayed onto the car body. A layer of a hard, clear lacquer could well follow this.

Special paints such as 'Hammerite' have been developed especially for metals. These paints do not require a primer or undercoat and can be applied directly to the metal surface. The finish achieved can be a smooth or hammered effect.

Powder coating

Powder coating is a method of applying paints to a product that has been statically charged. This is a dry process where a powder is used instead of paint. The powder is sprayed through an airgun. The powder is positively charged while the product is negatively charged, resulting in a very strong attraction between the coating material and the product. Once coated, the product is baked in an oven where the heat melts the powder over the product – producing a harder and tougher finish than is obtainable with conventional paints.

Examples of products coated in this way include domestic white goods such as fridges and washing machines, gates and fencing.

Since this is a dry process, it is environmentally sound – no solvents or propellants are used, and any excess coating materials can be recovered.

Example of a powder coated product

Plastic coating

Plastic dip coating can be used on a range of products. For example, a die-cast bottle opener, such as the one shown, has been dip coated on the main body and chrome-plated on the two levers.

The process involves heating the metal component to around 230 °C. The fine plastic granules are fluidised by passing air through them. This helps provide an even coating of material over the component being coated. The product is dipped into the fluidised polymer and removed. Heat from the product melts the plastic material, which cools in air providing an even coating over the product.

Task

For a child's swing, produced from mild steel, and intended for use outside, suggest at least two different applied barriers that could be applied to the material to protect it from the weather. Explain, using diagrams where appropriate, how these barriers would be applied.

Finishing processes for woods

Unprotected wood expands because it can absorb water – especially in outdoor applications. This causes the resins that bind the cells together to break down, rendering the material very weak. Bacteria and fungi can also attack wood.

Woods do not have the same natural protection against the environment as metals, and are prone to decay over a period of time. Hardwoods have a greater resistance to the environment than softwoods, due to the closer structure of the grain, and will therefore last longer. Teak contains oils that help repel rainwater and so protect the material.

Applied barriers

Paints

Paints can be used on woods, just as they can on metals, but first the material must be cleaned and any knots should be treated to prevent resin leaking. The surface of the wood must be sealed with a primer paint, which helps to key in the next layer of paint and prevents subsequent layers being absorbed by the wood. As with metals, an undercoat can be used followed by a topcoat. To obtain a high quality-finish, it is necessary to rub the surface down with fine sandpaper in between layers of paint.

Oil-based paints are generally used with woods. These are hardwearing, non-porous materials and are available in a range of colours. These paints can be used inside or out and are used for coating window and door frames.

Polyurethane paints, on the other hand, are used for coating children's toys, for example. They are extremely hardwearing, being tough and scratch-resistant.

If the natural grain of the wood is required, then sealing can be achieved by the use of transparent polyurethane varnishes. For products that are used outside, such as patio tables and chairs, a yacht varnish would be more appropriate. Yacht varnish is not affected by sunlight and the expansion or contraction of the timber in the same way that polyurethane varnish is.

Wood preservatives

The three main groups are:

- tar-oil derivatives (creosote);
- water-soluble preservatives;
- organic solvent preservatives.

All of these can be applied by brushing, dipping or spraying, or by pressure treating.

Creosote has been traditionally used for the preservation of a range of timber products. The main disadvantage with this method is that creosote will destroy any plants it makes contact with and is generally damaging to the environment. This has led to it being banned for use. An alternative would be the water-based solvents, which can be obtained in a variety of colours. As they are water-based, they have to be maintained more regularly.

Increasingly, timbers can be obtained that have been tannelised. This involves the impregnation of the timber with a solution of copper sulphate and other salts, which helps prevent water from being absorbed. Tannelised wood is pressure treated in large vessels. Water is first removed by drying, followed by filling the vessel with the preservative. Pressure is slowly increased until the desired amount of preservative has been injected into the timber. The wood is then steam dried before removing it from the pressure vessel.

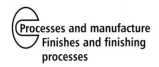
The main advantage of pressure treating timber is that it can be applied to uncut timber before manufacture of the finished product. Timber products treated in this way could well last up to fifty years.

Papers and cards

- **Varnishes** can be applied in the same way as ink; this is usually the final stage of the printing process and is carried out on the same printing machine. The aim of including this final stage is to extend the life of the printed product.
- **Laminating** is the application of a protective layer of a clear plastic material to the surface of the printed material. It is a separate finishing process, resulting in the protection of the printed material and, in addition, an increase in the strength and durability of the product.

Task

Explain how the two gates shown might be protected from the environment.

Steel gate   Wooden gate

PRODUCT ANALYSIS EXERCISE: *finishes and finishing processes*

Applied barriers

Study the photographs shown of:

(a) an anodised aluminium Mag-Lite torch;

(b) silver-plated cutlery;

(c) plastic-coated pliers.

1. For each of the products, use notes and diagrams to show how the finish is applied.

2. For each product, explain what function the finish has.

3. For one of the products, name an alternative finish. Explain why this finish is suitable.

PRODUCT ANALYSIS EXERCISE: *finishes and finishing processes*

Kitchen utility trolley

This kitchen utility trolley is made from solid beech.

1. Explain why this material is suitable for such an application.

2. The framework of the trolley is made from knock-down (KD) fittings.
   (a) Explain why KD fittings are often used in such products.
   (b) Use notes and diagrams to show two examples of KD fittings.

3. The drawers do not use KD fittings. Explain why this is so.

4. The wheels on the trolley are made from plywood. Explain why plywood is suitable for such an application.

5. Give a specific finish for the trolley. Explain how this is applied.

6. Explain what measures the manufacturers of this product would take to minimise impact on the environment.

Exam questions

AS exam question

1. (a) Name a specific finish that is used to prevent corrosion on ferrous metals. [2]
   (b) Describe how that finish is applied. [6]
   (c) Name a specific finish that is used to enhance the aesthetics of ferrous metals. [2]
   (d) Describe how the finish in part (c) is applied. [6]
   (e) Explain how ferrous metals corrode. [6]
   (f) Explain why non-ferrous metals do not corrode. [6]

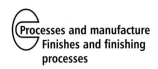

Processes and manufacture
Finishes and finishing
processes

A2 exam question

1. Metals can be finished with another metal to prevent corrosion, improve aesthetics or both.

   (a) With reference to a specific material, describe in detail a metal that is applied to another metal in order to prevent corrosion. [12]

   (b) With reference to a specific material, describe in detail a metal that is applied to finish another metal to enhance its aesthetics. [12]

Properties and materials testing

General properties of materials

All materials have properties to some degree or other. For example, a metal like aluminium will conduct electricity extremely well; where other metals like lead will conduct electricity, but not as well. A different material, e.g. a polymer such as ABS, will not conduct electricity at all, and is therefore an excellent insulator.

Materials' properties and structure

The properties of a material are determined largely by its structure. For example, metals are made up of crystals that contain atoms and molecules of the various elements that make up the material. Metals have good strength in both compression and tension, due to the very strong metallic bonds holding the atoms together. These metallic bonds also allow for free electrons to be shared amongst molecules, thereby making metals excellent conductors of heat and electricity.

Polymers and woods, on the other hand, have very different structures that do not permit the flow of electricity since there are no free electrons in their atomic structure. This makes them excellent insulators of both heat and electricity.

Although hardwoods and softwoods differ between type and structure, they are all fibrous materials made up of an arrangement of plant cells and resins. This results in the material having greater strength along the grain; woods are generally better in compression than in tension.

Polymers, in the main, are made up of long-chain molecules containing carbon, hydrogen (hence the term 'hydrocarbons') and oxygen atoms, along with other chemicals such as chlorine and fluorine.

The long-chain molecules in thermoplastics are held together by electrostatic bonds (called Van der Waals bonds). When heat is applied to the material, these bonds become weaker and so the material softens. With sufficient heat, the material can be remoulded. This is the basis under which thermoplastic polymers can be recycled.

The long-chain molecules in thermosetting polymers are held together by covalent bonds (very strong carbon-to-carbon bonds) forming rigid cross-links. These are not affected by heat, and so the material cannot be melted and reformed.

Materials' properties and products

One of the main factors affecting the choice of material for a product is the product's functional requirements; two examples are given below.

Example 1: A soft drinks can
The material for a soft drinks can:
- must not be affected by the acids in the liquid it is holding;
- must not contaminate the drink (i.e. it must be non-toxic);
- must have sufficient strength to withstand any internal pressure if the drink is carbonated;
- must be capable of being formed to the desired shape, e.g. one that is stackable in order to aid transportation. It must be capable of being deep drawn into the required shape. This means the material must have good ductility – the ability to be drawn out under tension without fracture.

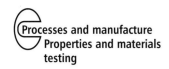
Processes and manufacture
Properties and materials
testing

When the can is being manufactured, the main body of the can starts life as a disc of material – in this case it would be mild steel. It is then formed into a cup shape, followed by deep drawing. This final process actually reduces the wall thickness of the can, resulting in the wall thickness being about one-third of the thickness of the base.

Example 2: An electric plug

Most domestic appliances are mains powered and so have a 13-amp plug fitted to them. If we consider the choice of materials for the main parts of the plug, i.e. the casing and the pins, we can see that they require quite different properties to function.

The casing

The casing is made from a polymer, either a thermoplastic or a thermosetting plastic. The functional requirement of this component is that it:

- acts as an electrical insulator;
- can withstand any heat generated by the flow of electricity;
- is sufficiently rigid and durable to withstand inserting and removing from a socket.

The pins

The pins, however, need to be able to conduct electricity and so are usually made from a metal. Durability is also a high priority for this component, since this is the part that actually moves against the contacts in the socket.

Definitions of materials' properties
Mechanical properties

Mechanical properties are those properties that determine how a material reacts to external forces.

- **Malleability** is the ability of a material to withstand deformation by compression. A malleable material, e.g. copper, can be deformed by compression before it shows signs of cracking. Malleability increases with temperature and therefore metals, which need to be bent, rolled or extruded, are heated first.

- **Ductility** is the ability of a material to be drawn out. Copper is also a ductile material, i.e. it can be deformed under tension before it fractures. The ductility of all materials decreases as the temperature increases, making them weaker at higher temperatures.

- **Toughness** is the ability of a material to withstand a sudden impact without fracture. It also refers to a material's ability to withstand bending (see also bending strength, below). Copper is a very tough material, e.g. copper wire can be bent many times before it fractures; while a high carbon (hard) steel possesses the opposite property – brittleness.

- **Elasticity** is the ability of a material to flex and bend when forces are applied and to regain normal shape and size when those forces are removed.

- **Plasticity** is the ability of a material to be permanently changed in shape by external forces, e.g. hammer blows, pressure, without cracking. Metals and (thermoplastic) polymers are generally more plastic when heated.

- **Hardness** is the ability of a material to resist abrasive wear, indentation or deformation. It is an important property of cutting tools, e.g. drills. Abrasive papers depend on the hardness of the abrasive medium to be effective. Brittleness is usually associated with hardness, unless the material structure has been altered to provide some measure of toughness, e.g. tempering of metals.

- **Durability** is the ability of a material to withstand wear and tear, weathering and the deterioration or corrosion this may cause. Weathering processes can change the appearance of a material and result in mechanical weakening.

- **Stability** is the ability of a material to resist changes in size and shape due to its environment. Timber tends to warp and twist due to changes in humidity. Metals and some plastics gradually deform when subjected to steady or continual stress. This gradual extension under load is known as 'creep'. Turbine blades are subjected to high temperatures and rotational speeds, therefore the blades need to be produced from a creep-resistant material.

- **Strength** is the ability of a material to withstand force without breaking or permanently bending. Different forces require different types of strength to resist them.
  - **Tensile strength** is the ability of a material to resist stretching or pulling forces.
  - **Compressive strength** is the ability of a material to withstand pushing forces which attempt to crush or shorten the material.
  - **Bending strength** is the ability of a material to withstand forces which attempt to bend the material.
  - **Shear strength** is the ability of a material to resist sliding forces acting against each other.
  - **Torsional strength** is the ability of a material to withstand twisting forces under torsion or torque.

Physical properties

Physical properties are those properties that refer to the actual matter that forms the material.

- **Fusibility** is the ability of a material to change into a molten or liquid state when heated to a certain temperature, i.e. a melting point. This varies between materials, but is essential to processes such as casting, welding and soldering.

- **Density** is defined as mass per unit volume. Relative density is the ratio of the density of the substance to that of pure water at a temperature of 4 °C.

- **Electrical conductivity:** all materials resist the flow of electricity to some extent. Electrical conductors offer a very low resistance to the flow of an electric current. Metals – especially silver, copper and gold – are good conductors. Liquids (electrolytes) and some gases also allow current to pass through them easily.

- **Electrical insulators** offer a high resistance to the flow of electricity. Non-metals are generally good insulators, but vary in their ability to resist the flow of electricity. Wood is a comparatively poor insulator, while ceramic materials, glass and plastics, such as nylon and PVC, are very good.

- **Semi-conductors** range between the two previous extremes, allowing electric current to flow only under certain conditions. For example, silicon and germanium in their pure state are poor conductors, but their electrical resistance can be altered by the addition of impurities.

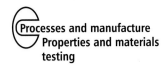
Processes and manufacture
Properties and materials
testing

- **Thermal properties** relate to a material's reaction to heat.
  - **Thermal conductivity** relates to how heat travels through the material. It is measured in watts per metre per degree Celsius. Metals, especially copper and aluminium, possess high thermal conductivity.
  - **Thermal insulators** are materials with low value thermal conductivity, and are generally non-metals. They are used to prevent heat gains and losses, e.g. pan handles, loft insulation.
  - **Thermal expansion:** materials generally expand when they get hot and shrink upon cooling. A material's coefficient of linear expansion is the fractional change in length due to changes in temperature. In large civil engineering projects, such as bridges, allowance has been made for movement caused by seasonal variations of temperature. Control mechanisms also use this effect, e.g. car thermostats or automatic kettles.
- **Optical properties** refer to how materials react to light and heat by reflection, radiation and absorption. This will vary according to whether the material is translucent, transparent or opaque:
  - **Opaque** does not allow light to pass through, e.g. house brick, mild steel sheet.
  - **Translucent** allows some light to pass through, e.g. one sheet of photocopy paper, a thin acrylic sheet, fine bone china.
  - **Transparent** allows light to pass through, e.g. clear glass.
- **Magnetism** is a property possessed by many elements, and is the product of the orientation of electrons about their atoms. Only a few elements, however, are strongly magnetic.

Materials testing

At AS level
As an AS level student you need to have an understanding of basic tests that can be applied to materials to identify a material, and its suitability for the intended application. All of the tests suggested here can be applied easily within the workshop using the equipment usually found there.

At A2 level
A2 students should build on their knowledge of basic testing and begin to appreciate the testing of materials and products in an industrial environment.

Workshop testing
What kind of material is it?

Metals
Metals are readily identified by their density, colour and shine (or the way the light reacts with it). By simply looking at the material and judging these aspects, a metal can be identified. For example, aluminium is silvery in colour but is much lighter than silver. Polished steel can be similarly silver in colour, but is also heavier than aluminium.

Aluminium is a soft metal, so if you were to scratch it with a file or scriber it would make a deeper mark than mild steel. If all else fails, you can test the material with a magnet. Mild, medium and high carbon steels are all magnetic. Stainless steels and most other metals are not.

Task

Some metals have generally the same appearance; think about how you might identify steels among a collection of metals.

Woods

Woods can be identified by their grain, colour and density. Beech (a hardwood) has a very close straight grain with very few knots and is a light buff colour, while Scots pine (a softwood) also has a straight grain but is very knotty. Some woods have a distinctive grain such as walnut or yew (both hardwoods), while woods like oak, as well as having a distinctive grain, are heavy.

Plastics

Plastics can be more difficult to identify simply by looking at them and, thankfully, nowadays manufacturers employ an identification code to the materials. This enables the material to be collected and recycled. However, there are aspects of plastics' properties that can help identify the material.

- Determining whether a plastic is a thermoplastic or thermosetting material can be done by the application of heat; if it begins to go soft, then it is a thermoplastic.
- Polyethylene, for example, will float.
- Acrylics and polystyrenes can shatter on impact.

Is the material appropriate?

Workshop testing of metals can be carried out for:

- hardness;
- ductility and malleability;
- tensile strength;
- toughness.

Hardness

Hardness can be tested in a number of ways – two are given here. (Both of the following methods can also be used to compare two materials.)

- A file can be run over the material. If the file cuts, then the material is soft (or rather softer than the file); if it does not cut, then it is harder than the file.

- A dot punch is used to create an indent in the material. It is most important that when comparing materials, the amount of effort used in striking the dot punch with the hammer is the same for all sample pieces.

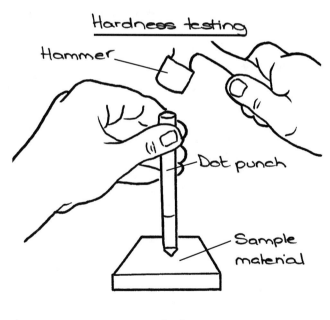

Hardness testing

Hammer

Dot punch

Sample material

Indent produced by dot punch

Lead

Copper

Mild steel

Processes and manufacture
Properties and materials
testing

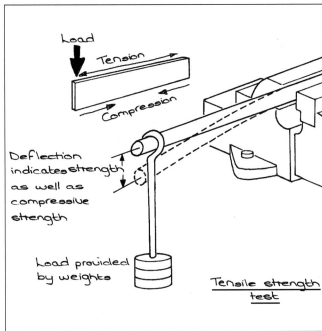

Cracks along the outside of the bend indicate low ductility

Testing for ductility

Load

Tension

Compression

Deflection indicates strength as well as compressive strength

Load provided by weights

Tensile strength test

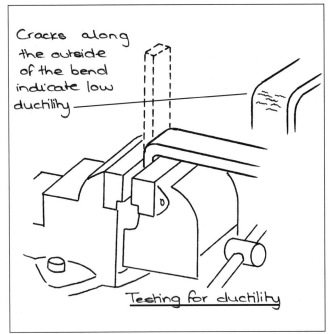

Testing for toughness

Brittle ⟷ Tough

Test piece in vice hit by hammer

WARNING: Safety spectacles must be worn.
Some materials may shatter.

Ductility and malleability

Ductility can be tested by placing a piece of the material in the vice, then attempting to create a 90° bend. If the material shows cracking on the outside of the bend, then it may not be sufficiently ductile for its intended purpose. If cracking appears on the underside of the bend, then the material may not be sufficiently malleable.

Tensile strength

An indication of tensile strength is the amount of energy required to bend a material. A comparison of tensile strength can be made by clamping sample materials of similar length in the jaws of a vice, and applying the same load to them. In this case, it is the resistance of the material to the load being applied that is being tested.

Toughness

Toughness is a material's ability to absorb mechanical shock from, for example, a hammer blow. A comparison of materials' toughness can be made by clamping samples of materials in a vice, then striking them with a hammer.

Task

Take a variety of materials – metals and plastics – and carry out the four tests above. Try to put them in rank order of hardness, ductility and malleability, tensile strength, and toughness.

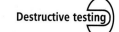

Note Take care when testing plastics: some may be brittle and shatter quite easily. Safety glasses should be worn when carrying out these tests.

Industrial testing

Why is it necessary to test materials? The most obvious answer is that the designer must know that a material is suitable for its intended purpose, i.e. whether or not it has sufficient strength, toughness, durability, etc.

In the previous section, we looked at simple workshop tests that helped us decide what type a material was and provided some indication of strength, hardness, toughness and ductility. In this section, we will look at standard methods of testing materials and products that are used throughout a range of industries. Materials tested using these methods are provided with certificates stating their specific materials' properties as numeric values.

Destructive testing

Destructive testing refers to a range of tests that ultimately results in the destruction, usually, of a standard-size test piece. Test pieces are of a standard size and shape for the particular test being carried out.

Hardness test pieces will have small indentations after testing, while impact testing equipment will cause the test piece to bend or break completely. Destructive testing also refers to tests that are applied to products, for example, testing for durability. Tyre companies such as Michelin will run tyres on vehicles to establish wear-rates in use.

Hardness tests

Hardness is the ability of a material to resist abrasive wear, indentation or deformation. Hardness is actually a by-product of strength – generally, the stronger the material the harder it is. In order to establish hardness of a material, it must be indented or deformed.

There are three basic methods of hardness testing:

- the Brinell test;
- the Vickers test;
- the Rockwell test.

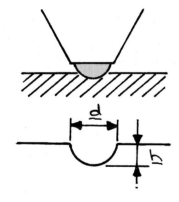

Brinell hardness test

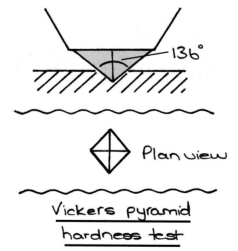

Vickers pyramid hardness test

A hardened steel ball indentor is used. Diameter **d** and depth **h** are measured and used to calculate a hardness value.

A diamond pyramid indentor is used. The surface area of the indent and the load are used to calculate a hardness value.

Processes and manufacture
Properties and materials
testing

All tests produce an indentation of the material or product. The Brinell and Vickers tests produce an indent in the material being tested that has to be measured with the aid of a microscope.

- In the Brinell test, a hardened steel ball is forced into the material's surface by means of a suitable load. The resulting surface area of the indent is measured, and used to calculate the hardness number.
- The Vickers test uses a diamond pyramid to indent the material. This is again measured using a microscope to give the hardness value.
- The Rockwell test is more appropriate for the quality control testing of finished products. It is a relatively rapid test, with the hardness value being indicated on an attached dial – thus avoiding the need to measure a very small indentation.

Different scales can be used, depending on the material being tested.

Tensile tests

Tensile testing involves putting material under tension by stretching to provide information regarding tensile strength, elasticity and plastic properties, such as ductility and malleability.

Essentially a standard test piece is held between two grips. One of the grips is fixed, while the other is attached to a vertical slide operated by electric motor. The test piece is then put under tension at a constant rate until (i) it breaks, or (ii) it stretches beyond the limits of the machine.

As the test piece is being stretched the distance travelled by the vertical slide is recorded and plotted against the load being applied, which is sensed by a load transducer fixed to the moving grip. This will result in a graph that can be used to indicate a variety of useful information (see page 123).

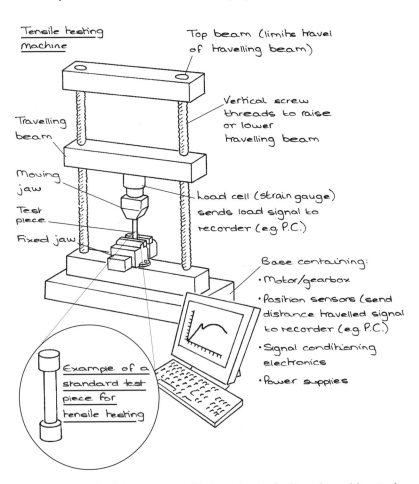

Tensile testing machine

Top beam (limits travel of travelling beam)

Vertical screw threads to raise or lower travelling beam

Travelling beam

Moving jaw

Test piece

Fixed jaw

Load cell (strain gauge) sends load signal to recorder (e.g. P.C.)

Base containing:
- Motor/gearbox
- Position sensors (send distance travelled signal to recorder (e.g. P.C.)
- Signal conditioning electronics
- Power supplies

Example of a standard test piece for tensile testing

Tensile testing is standard in a range of industries including the rubber industry, where the rubber, steel and fabric used in the manufacture of tyres are tested. This method of testing is also used in the rope-making industry to test the strength of the fibres being used, and in the clothing industry to test the strength of fabrics.

Tensile test diagram for mild steel

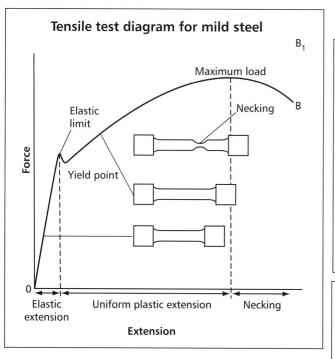

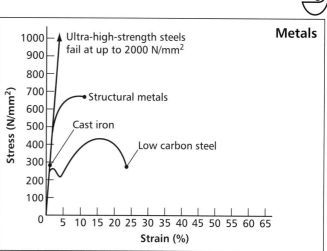

Metals

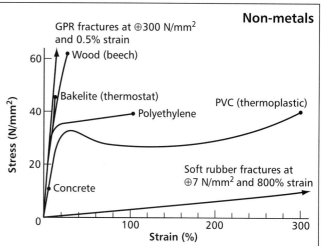

Non-metals

Examples of tensile testing results

Task

Suggest reasons why standard shaped/sized test pieces are essential when carrying out these types of tests.

Impact tests

Impact tests indicate the toughness of a material and, in particular, its resistance to mechanical shock. Toughness, or rather its opposite property of brittleness, is often not revealed by a tensile test.

There are three main methods of impact testing:

- the Izod test;
- the Charpy test;
- the Houndsfield test.

Test pieces have been standardised for each test and are shown below. The more the sample piece absorbs the impact, the tougher the material is.

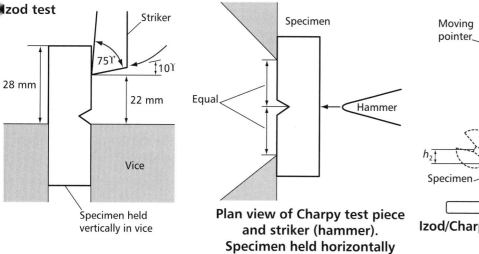

Izod test

Specimen held vertically in vice

**Plan view of Charpy test piece and striker (hammer).
Specimen held horizontally**

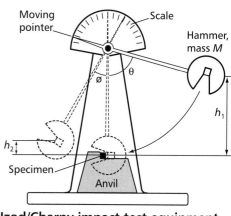

Izod/Charpy impact test equipment

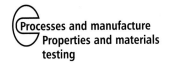
Processes and manufacture
Properties and materials
testing

Wear resistance

Wear resistance of a material can be determined by rubbing with an abrasive. A standard test involves weighing a disc of material, e.g. for use in brake discs, rubbing an abrasive disc against the material a given number of times, at a given pressure, followed by re-weighing to give a quantitative value of abrasive wear.

In the pottery industry a similar test is applied but in this case the abrasive resistance of the glaze applied to crockery is tested using an object similar to a dinner knife.

Fatigue testing

As with most tests both materials and products can be tested for fatigue. For example, the wire used in tyres is tested by gripping a sample between two chucks.

One chuck is stationary while the other can rotate – being driven by a motor/gearbox. The sample is twisted by rotating one of the chucks for a given number of revolutions. The direction of rotation is changed, and so on, until the sample breaks.

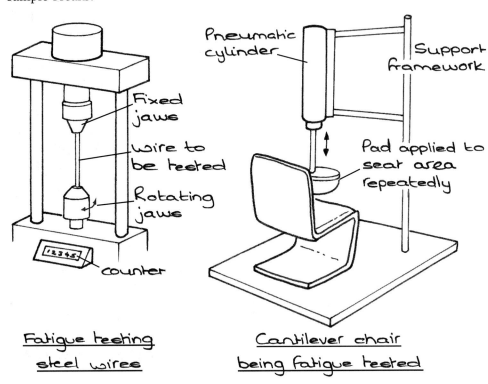

Sketch showing an example of a fatigue test on a chair

Products such as chairs can be tested for fatigue. In this case a weight (or force) is applied repeatedly to the product. The diagram shows a chair being tested by the application of a force on the seating area. The force being applied is representative of a person sitting on the chair and is used to test the resistance of material.

Non-destructive testing

Non-destructive tests, as the name implies, are methods of testing that do not damage or destroy the material or product. These methods of testing are usually carried out on the final product, where the aim is to test for surface or near-surface faults or flaws. There are a number of standard methods of testing that can be found in a range of industries, and some methods require the use of sophisticated equipment. Typically, welded joints and castings are tested in this manner.

Surface crack detection

Sound and touch

These are methods of testing that can be applied to a variety of products and use basic quality-control techniques.

- 'Ringing' – just as the term suggests, a bell cannot ring if it is cracked. This method of testing a product is used in both the casting and pottery industries, where ringing is the first line in detecting a faulty product.
- As well as a visual check, touch can also be used to sense faults directly on the surface of the product.

Liquid penetrant

- The penetrant liquid is sprayed onto the surface of the product.
- Any excess is removed leaving only the penetrant in the cracks.
- To help make the cracks clearer, a light dusting of chalk is applied. The coloured dye marks the position of the crack in the white chalk.

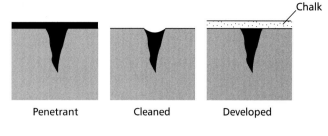

Chalk

Penetrant   Cleaned   Developed

Crack detection

Magnetic testing

The component is magnetised by making it part of an electromagnetic circuit. Iron particles are dusted over the area, which then highlight where the magnetic lines of force are broken by a defect. This method is particularly useful for finding defects just below the surface.

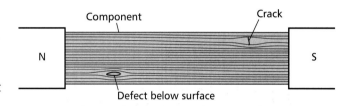

Component   Crack

N   S

Defect below surface

Component within a magnetic circuit

Acid pickling

In order for cracks to be visible it is necessary to clean the surface of the product, such as casting. Steel castings, for example, can be 'pickled' in a weak solution of sulphuric acid, heated to 50 °C, for a number of hours. This will remove any oxides from the surface of the material, and when washed will make surface cracks clearer.

Internal defect detection

Light

Light is utilised in the pottery industry to 'see' through some of the finer ceramics, such as bone china, to indicate faults within the body of the material.

X-ray methods

X-rays are used to detect defects under the surface. For example, in the tyre industry X-rays are used to check for air bubbles between layers of rubbers, fabric and supporting wires. An X-ray tube emits radiation through the product and forms an image on a photographic plate, or through an image intensifier and camera, to a monitor. X-rays travel faster through cavities and so produce a darker image where the structural materials have not vulcanised together properly, which would have resulted in a potentially dangerous product. (See diagram overleaf.)

Example of an X-ray arrangement

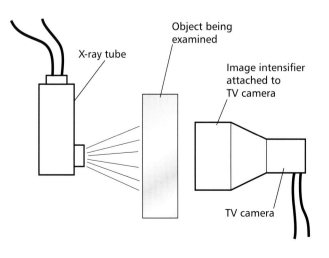

Object being examined

X-ray tube

Image intensifier attached to TV camera

TV camera

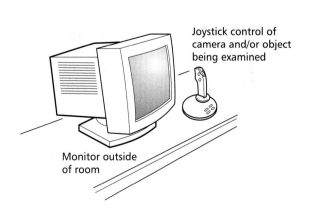

Joystick control of camera and/or object being examined

Monitor outside of room

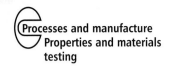

Processes and manufacture
Properties and materials
testing

γ-ray methods (gamma rays)

Used in the same way as X-rays but since γ-rays are 'harder', thicker and denser, materials such as cast iron and steel can be tested to detect 'air' pockets in the casting.

Both X-rays and γ-rays are lethal to human tissue. γ-rays are produced from strontium-90, present in the 'fall-out' products of nuclear explosions. For this reason, such methods are shielded behind puddle (dense) concrete or lead-lined walls. Also equipment operators wear a photographic film badge, which is monitored to indicate level of exposure to the X- or γ-rays.

Ultrasonic testing

Very high frequency sound vibrations can be used to precisely locate internal defects. Ultrasonic frequencies between 500 KHz and 100 MHz are used. A probe is passed over the component transmitting the high frequencies. Under normal conditions, the vibrations will pass through the material and will be reflected back from the bottom surface of the material. The probe receives this and an amplifier converts the vibrations into a series of blips on a monitor.

Ultrasound is used to test sheet, plate and strip materials up to 6 mm thick. It is therefore ideal for tyre, aircraft and pipeline inspection in pinpointing defects.

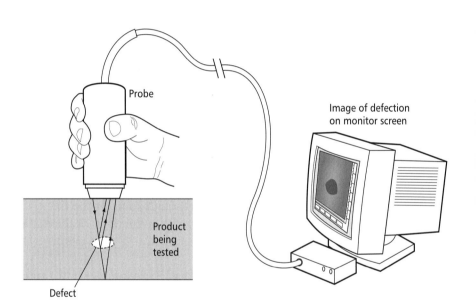

Ultrasonic testing

Steel and alloys used in industry

When products are manufactured, the materials used have to be tested to ensure consistent quality. In the manufacture of cars, the quality of the steel is vital. Specific steel is selected for its mechanical properties, such as malleability and ductility. These properties are crucial in order to press body panels. The steel needs to be malleable so that it can be formed by pressing methods without the steel tearing or cracking. Ductility is needed so that the steel will stretch over the pressing dies.

At Swindon Pressings (part of the BMW group), the body panels for the Mini, Land Rover Freelander and other MG Rover vehicles are manufactured. Test engineers will perform random materials testing on the steel that they use, because this is imported from outside suppliers. It is very important that the steel used matches the original design specification. Tests might include tensile tests to check the point at which the material fails under pulling forces. This will give an indication of how ductile the material is. The metal may also undergo impact testing, where samples of the material are snapped under impact and the structure of the steel examined. Samples will be examined under a microscope, where a fibrous structure will indicate a ductile material. Such materials testing is known as destructive testing as the sample is damaged as a result of the test.

Following such materials testing, test pressings will be done to assess how the steel performs. These will be checked against 'standard' pressings located in each pressing area.

In the manufacture of products such as alloy wheels, or the undercarriages of aircraft also made from lightweight alloys, non-destructive tests might be used to ensure quality. Products such as these are made by casting methods. Castings can be prone to internal cavities invisible to the naked eye. In order to test for such defects, X-rays are used. Internal cavities will normally show up as a dark part on X-rays. Hairline surface cracks may also be very difficult to see. These are tested by painting 'engineer's blue' onto the component while the component is hot. The heat evaporates the spirit base of the engineer's blue, and the blue dye will concentrate where there are surface defects.

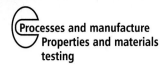
Processes and manufacture
Properties and materials
testing

Exam questions

AS exam question

1. (a) With reference to specific materials, describe simple workshop tests that you could do to test for the following properties:
   (i) malleability;
   (ii) toughness;
   (iii) hardness. [3 × 7]
   (b) When selecting woods to make a product with, what checks would you make in order to ensure quality? [7]

A2 exam question

1. A manufacturer of children's toys needs to test materials for their safety. With reference to specific materials, explain in detail what tests may be carried out. [24]

Modern manufacturing systems

Introduction

In industry today, a number of systems are used to ensure that customers get the products they want, on time and to the correct level of quality.

At AS level
As an AS level student you should be familiar with the general systems used.

At A2 level
A2 students should be familiar with manufacturing systems used in some specific industries.

Quick response manufacturing (QRM)

In today's market, consumer demand changes rapidly with changes in fashion. This means that many industries can no longer manufacture items to go into stock or storage before sale. Companies that do manufacture in this way run the risk of not being able to sell their products if there is a sudden change in market demand. Instead, manufacturers today often prefer to 'make to order'. For example, at Jaguar Cars, every vehicle made has a customer before its assembly is started. One immediate advantage of this is that it avoids expensive storage costs while waiting for the car to be sold. The main disadvantage of manufacturing in this way is that the customer may have to wait longer for the product. The use of QRM systems reduces this waiting time.

Electronic point of sale (EPOS)

Electronic point of sale refers to the technology of using bar-coded products that are laser scanned at the point of sale. This system enables the sale of an item to be registered with distributors, warehouses, etc., who in turn can electronically re-order stock from the manufacturer. EPOS, therefore, enables manufacturers to produce items 'just in time' (JIT), rather than storing them. This is an essential part of QRM.

Bar codes or 'data communication tags' are not only used at the point of sale, but also in the identification of a component or a finished product while it is being processed or stored.

Just in time (JIT)

'Just in time' is a system devised to ensure customers get the products they want at the right time, and to ensure that manufacturers do not have to stockpile raw materials or components.

Using JIT, manufacturers organise their suppliers to deliver materials and components just in time for when they will be needed for production. This avoids the need to have large amounts of stock cluttering up assembly lines or the need for costly storage. In practice, companies such as Jaguar may have enough stock to make the cars planned for assembly in one shift.

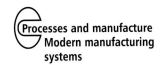

JIT places the responsibility of ensuring parts and materials arrive on time with the supplier. In order to supply companies such as Jaguar, manufacturers would have to guarantee not only high quality at the right price, but also that supplies arrive exactly on time.

Kanbans

The control of stock levels is managed using computer systems. Many parts or components that are used in the manufacture of cars are delivered in containers, and then placed at the work cell they are needed in. Each container is labelled with a bar code containing information to identify both (i) the part, and (ii) the quantity contained in the box. Inside the container, a card is kept with the same information on. This is known as a 'kanban' (Japanese for 'card signal'). As the operator working in a cell starts to use up the items in the container, the kanban is placed in a chute, where it will go into a small box for later collection by the store's workers. When the kanbans are collected, they are scanned and, using electronic data interchange (EDI), the parts can be automatically re-ordered from the supplier.

Although the term 'kanban' is, strictly speaking, the card system described above, it is also a term generally used to describe a computer system that is used to control the flow of products and components through a system.

Sequencing

An essential part of a JIT system is the sequencing of work. Once parts or materials have arrived at a factory, they need to go to the individual work cells at the right time. This is controlled, again, by the use of computers.

Master production schedule (MPS)

MPS is a computer-controlled scheduling system that sets the quantity of each product to be made in a given time period. In the car industry, this is done using 'order-based' scheduling, because cars are made to individual order. (Customers choose body colour, individual accessory packs, interior finish, etc.) In this system, cars would be assembled in order priority. In planning this, materials requirements planning (MRP) software is used to order the required materials and components for each vehicle.

Telematics

Telematics is a system used to electronically track a product from receipt of customer order through to assembly and dispatch. In the car industry, customer orders are converted into electronic data programmed into a 'black box'. This is placed on the car as it goes through each part of assembly. The progress of the order can then be tracked at monitoring stations, and operators can check that components they fit, such as engine type, stereo options and so on, match the specific customer order.

Flexible manufacturing systems (FMS)

Manufacturers who mass-produce items such as aerosol cans, toothpaste tubes and so on, will use dedicated, automated equipment that will only produce those items. Investment in such production systems is relatively 'safe', because the demand for such mass-produced items is fairly constant. In other markets, items may have to be made in batches. In order to do this, the equipment used needs to be more flexible.

The following equipment can be made to be flexible:

- press formers;
- CNC (computer numerically controlled) punches;
- CNC laser cutters;
- CNC lathes and milling machines;
- robot arms.

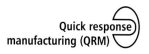

Press formers

Press-forming equipment can be flexible, because it is possible to change the dies so that different products can be made. Obviously, this is limited to the number of dies available.

Car manufacturers will typically change press dies so that they can make more than one model of car on the same assembly line. For example, at the Jaguar Halewood factory, the X-type Jaguar and the X-type estate are made on the same line.

CNC punches

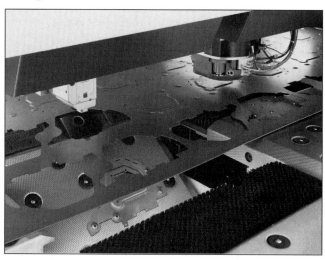

CNC punching

Computer-controlled punches can be programmed to punch a variety of designs out of sheet metals. Usually, they have a magazine of tools that can be changed automatically as required for the work they are doing.

CNC laser cutters

Computer-controlled lasers and flame cutters can be programmed to cut out a wide variety profiles, slots, holes, etc.

JCB uses CNC lasers to cut the steel used to build vehicle chassis. As different models are made, it is necessary to be able to quickly download new programs to the same machines.

CNC lathes and milling machines

Lathes and milling machines can be programmed to do highly accurate one-off jobs or batches of items.

Typically, lathes are used to turn the diameter of bars, machine screw-threads, and face and drill bar ends. Milling machines can be used to cut slots, pockets, drill holes or cut profiles. The use of CAD/CAM (computer-aided design/computer-aided manufacture) software has made re-programming such machines relatively easy.

Robot arms

Robots can be programmed to do:
- the same job all of the time on one product;
- the same job, but on different products;
- entirely different jobs.

They can be provided with different tools, or end effectors, so that they can do tasks such as spot welding, spray painting, applying adhesives, lifting components, and so on. Again, re-programming is relatively easy, and so such robots are quite flexible.

Robot arms used to spot weld Jaguar cars can have alternative programs, so that they can weld the different-shaped body or chassis panels according to the model that is being made.

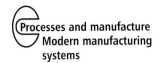
FMS cells

Plan view of a typical FMS cell (flexible manufacturing cell)

Material in

CNC turning

1

2

Buffer store

3

Parts handling robot

Control unit

4

CNC milling

Finished components out

Flexible manufacturing machines can be organised into cells. In addition to the machines, they may also include a buffer store. This is a temporary store used to hold partially completed work. As some machines will work faster than others, and some jobs will take longer, work that is partially completed will need to be held while the other jobs take place.

For an FMS cell to work, a host computer is needed to control the sequencing of jobs, and to monitor the status and performance of each machine in the cell. This host computer will download CNC part programs according to customer orders. Because the machines are flexible it is possible to have several different programs running and, therefore, different products can be made in the same cell at any one time.

Task

1. Discuss what is meant by the term 'part programming'.

2. Discuss why it is viable to use CNC machines for both one-off and volume production.

3. List the advantages of using flexible manufacturing systems.

Further reading

- *Fundamentals of Modern Manufacturing: Materials, Processes and Systems, Second Edition*, Mikell P. Groover (Wiley Text Books)

- **www.autoindustry.co.uk** for a very useful dictionary of manufacturing terminology

PRODUCT ANALYSIS EXERCISE: *modern manufacturing systems*

Shampoo bottle

1. Name a material that is suitable for the manufacture of the bottle body. Explain why this material is suitable.

2. Use notes and diagrams to show how the bottle body would be manufactured.

3. Name a material that is suitable for the manufacture of the bottle top. Explain why this material is suitable.

4. Use notes and diagrams to show how the bottle top is manufactured.

Exam questions
AS exam questions

1. Explain what is meant by the term 'just in time' (JIT).　　　[5]

2. Explain how data communication tags or bar codes can be used to control stock levels.　　　[5]

3. Describe the advantages and disadvantages of using flexible manufacturing systems.　　　[8]

4. Use notes and diagrams to show how a flexible manufacturing cell could be organised.　　　[10]

A2 exam questions

1. Manufacturers today often prefer to make to order rather than produce goods to stock. This is done by using quick response manufacturing (QRM). With reference to a specific industry that you are familiar with, describe how companies achieve QRM.　　　[24]

2. Flexible manufacturing systems are replacing traditional methods of manufacture. With reference to the manufacture of a specific product, explain:
   (a) how flexible manufacturing is used to manufacture the product;　[12]
   (b) the benefits of using flexible manufacturing systems.　　　[12]

Computer-aided manufacturing

Introduction

At AS level

As an AS level student you should be familiar with the common types of computer-controlled equipment used in industry, and be aware of the benefits that the use of CAM has. You should also be aware of the broad types of robot used and have understanding of the benefits they bring to manufacturing.

At A2 level

A2 students should have a good knowledge of some specific industrial examples of using CAM, and be able to present a balanced argument for the advantages and disadvantages of its use. They should also be able to describe how robots are used in specific industries, and have an insight into the benefits and drawbacks of using robots.

CAM and product development

In order to speed up the design and development of products, computer numerically controlled machines, such as five-axis milling machines, can be used to machine moulds or for tooling directly from the data generated from a three-dimensional CAD drawing.

Example 1: Wedgwood

Wedgwood uses computer-aided design and manufacturing software (CAD/CAM) to convert three-dimensional CAD drawings of its designs into machining data. This data is then used to machine 'blocks', which are highly accurate master moulds. The blocks are then used to make plaster moulds, which will later be filled with clay slip, in the production of hollow ware. Again, the use of CAD/CAM greatly reduces product development time, because it would take a skilled craftsman considerably longer to make a 'block' by hand.

Example 2: JCB

In other industries, computer-aided manufacture may be used to process materials. For example, at JCB Bamford, Excavators, CNC lathes are used to machine highly accurate steel pivot pins (used in the articulating joints of the back hoe). In the production of the vehicle chassis for back-hoe excavators, JCB uses CNC laser cutters to precision cut steel plate. Following this, the steel plate is pressed and folded into shape using computer-controlled hydraulic presses. It would also use computer-controlled robots to carry out precision welding operations, such as in assembling parts to the chassis.

Three-dimensional scanners

Three-dimensional scanners are scanners used to measure an object by tracing a series of points over the surface of the object. The measurements are then used to build up a three-dimensional digital map of the item. This data can then be imported into CAD software and edited to make three-dimensional rendered drawings.

At Wedgwood, three-dimensional scanners are used to scan ceramic objects stored in the company's archives. These objects might well be over two hundred years old, but may have designs, engravings and so on that can be used on modern-day products. Because they can be digitally captured, it is possible to make reverse images or patterns, 'emboss' a design below the surface, or create a relief image to any size, all from one scanned item.

Wedgwood uses two types of three-dimensional scanner.

- **Contact scanner:** this is large, using a probe that makes physical contact with the object. The probe is driven by a CNC program, which ensures that the object being scanned is measured precisely. Such scanners are normally used for fairly large items and can take some time to scan an object.

- **Non-contact scanners:** the most common non-contact scanners use lasers to measure an object. Wedgwood uses this type of scanner to scan very detailed (often small) items, such as those that might have an engraved or relief pattern. Laser scanners are extremely accurate and very fast. As they do not make contact with the item being scanned, there is also no risk of damaging the surface.

Task

With reference to other industries you may be familiar with, write a bullet-point list of the benefits of using computer-aided manufacture.

Robotics

Categories of robot

The categories are:

- first generation;
- second generation;
- third generation.

First generation

This type of robot responds to a pre-set program and will carry on regardless of any external changes. For example, if the robot was packing eggs into a carton, and one egg broke, the robot wouldn't know and would just continue packing. This type of robot is becoming obsolete, as it has limited use in modern industries.

Second generation

This type of robot is fitted with sensors, which are used to feedback information to a central control computer. This information is then used to monitor the operation of the robot and to automate the work cell.

For example, robots placing steel blanks into a press would firstly collect the blanks from a pallet. Sensors would need to be used:

- to ensure that the pallet is loaded with blanks;
- to ensure that the blank is being presented into the press the correct way around;
- as the pressing is removed, to check that it has been pressed correctly.

The most common sensing method today in robotics is the use of digital cameras. Pictures of components being worked on can be automatically compared to reference pictures stored on the host computer. The host computer can then automatically stop a work cell, if it detects an error.

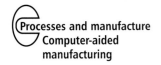

Third generation

Third-generation robots use sophisticated sensors and computer programming to create 'artificial intelligence' or AI. This type of robot is able to not only detect changes to the environment that it is working in, but also modify its own program and therefore its actions in response to the changes. So far, AI robots are still generally only in experimental stages but, for industry, there is the potential to develop robots that can detect faults, diagnose the problem and rectify it.

Robot configurations

There are three main types of robot used in industry today:

- beam transfer;
- arm;
- automatic guided vehicle (AGV).

Beam transfer robot

Beam transfer

These are relatively simple robots that operate on parallel slides or beams. The robot will move along x- and y-axes, and are generally used to pick up a component from one machine or a pallet and place it in another. This type of robot is sometimes called a 'pick and place' system.

Beam transfer robots can be seen in the car industry, where they are used to pick up press-formed body panels and move them along the manufacturing line.

Arm

This type of robot configuration is the most versatile. Robot arms are jointed in a similar way to the human arm, having a shoulder, elbow and wrist. These joints and the directions they can move in are known as 'degrees of freedom'. The more degrees of freedom a robot has, the more useful it is.

The hand of a robot arm is known as the 'end effecter'. This can be fitted with a wide range of tools, including:

- air guns for spraying;
- spot welders;
- laser or flame cutters;
- manipulators (often pneumatic suction cups) for picking items up.

Automatic guided vehicle

Automatic guided vehicles

Automatic guided vehicles, or AGVs, are rather like a fork-lift truck without a driver. They are robots that are used to carry components and finished items around factories.

- AGVs navigate by either using sensors, which follow a wire that is buried under or stuck to the surface of a factory floor, or by using lasers, which bounce off reflectors placed high up on walls. In the latter, the robot takes three measurements at any one time, and can triangulate its position according to a map of the factory layout stored in its memory.
- AGVs are often programmed to interface with factory 'just in time' (JIT) systems, so that they deliver materials and components to the right place and at the right time.

Programming methods

There are three main ways to program robots.

- **Teach pendant** is similar to using a remote control, where an operator will use the control or teach pendant to guide the robot through a series of movements. The control computer stores and converts the movements into a control program.

- **Walkthrough** is where an operator will physically pull the robot through the required movements, whilst the control computer again records and converts these movements into a control program. This type of programming is useful to 'train' robots in operations such as welding or spray painting.

- **Off-line** is one of the most popular ways to program robots. Virtual reality simulations of a work cell can be used to program a robot and to test the program, without the risk of damaging the robot or anything else in the work cell. They can also be used to rehearse dangerous operations, such as maintenance tasks in the nuclear power industry.

The benefits of using robots

Robots can:

- carry out mundane, repetitive tasks that humans dislike, e.g. loading or unloading components from machines;

- carry out physically demanding jobs where there might be a risk of repetitive strain injury, for example when lifting and moving heavy components;

- work in hazardous areas, such as in work cells where spot welding, arc welding, laser cutting, spraying, is taking place; or in the nuclear industry, to carry out inspection and maintenance on radioactive equipment;

- work to high levels of accuracy, consistently and quickly. For example, in spot welding car body panels, robots will place the correct number of welds in the correct place, every time. Human operators cannot work consistently with such accuracy while having to work at speed;

- work for long periods of time without the need to stop (apart from maintenance). Maintenance stops can be programmed in to avoid machine breakdown or faults on the product occurring. For example, at MG Rover, robot spot welders will automatically clean their copper electrodes every few welding cycles. The electrodes will then adjust to compensate for any copper that wears away.

The drawbacks of using robots

There are several problems associated with using robots. They are:

- poor mobility and flexibility;
- limited degrees of freedom;
- high set-up costs;
- employment issues.

Mobility and flexibility

Robots have poor mobility and flexibility compared to human workers. A human worker can move easily from one manufacturing cell to another and work on entirely different tasks, if required. Robots, on the other hand, can be difficult to relocate because of their size and, in order to carry out different tasks, they would have to be re-programmed and often re-tooled. Humans can pick up different components and different tools easily, whereas a robot's end effecter tends to be dedicated to doing one or two tasks.

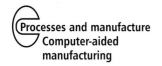

Degrees of freedom

Humans can work in tight spaces and move in and out of those spaces relatively easily. For example, when installing dashboard or steering wheel assemblies, humans can do the job much more easily than robots.

Set-up costs

Robotic cells are firstly very expensive to purchase and, depending on the tasks they are to perform, can be hugely expensive to program and set up so that they function correctly.

Employment issues

Robots are often used to replace labour and, therefore, there can be some loss of jobs. This can lead to poor labour relations, and so on. In addition to this, employees need to be able to adapt to working with the new technology. They need to be willing to train on the use of such technology and be flexible to work in, perhaps, different ways (often taking responsibility to supervise work cells and problem-solve on their own initiative).

Task

Using the Internet, research two different types of robot used in industry. Make notes on how the two robots are used in specific companies.

Further reading
- www.21stcentury.co.uk/robotics
- www.kawasakirobot.uk.com

PRODUCT ANALYSIS EXERCISE: *computer-aided manufacturing*

Study the photographs of the engineering components shown and answer the following questions.

1. The engine part shown has been manufactured using CAM. Name the type of machine that could have been used.

2. Explain why the part would have been manufactured using CAM.

3. What are the disadvantages of using CAM in the manufacture of products?

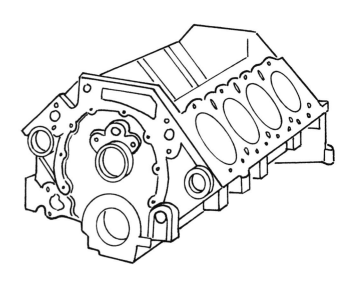

Engine block

Exam questions

AS exam question

1. (a) With reference to a material area of your choice, describe how designers might use CAD in the design and development of products. [10]

   (b) Using **two** different examples from your chosen material area, describe how rapid prototyping is used in the development of products. [2 × 5]

A2 exam questions

1. Using examples that you are familiar with, describe how industry uses ICT in the design, development and manufacture of products. Your answer should make reference to the benefits for the manufacturer in using ICT. [24]

2. (a) With reference to specific industrial examples, describe how robotics are used in the manufacturing process. [12]

   (b) Explain what impacts the use of CAM in industry has on employment. [12]

ICT in manufacturing

Introduction

The use of computers in manufacturing is a subject that all AS and A2 students should investigate.

At AS level
As an AS level student you will need to be familiar with the broad uses of ICT within industry and the advantages it brings.

At A2 level
A2 students should have a thorough understanding of the use of ICT in industry, and be able to recall some specific examples of how and why ICT is used.

Scales of production

Products are made in a range of quantities: from a one-off to large-scale production runs, depending on demand. The numbers produced indicate the scale of production for a product.

An example of a piece of bespoke jewellery

One-offs

One-offs refer to those products that have been designed and manufactured for a single specific situation. For example, a pair of wrought-iron gates specially designed for the gap they are to fill, or a custom-made bicycle frame for a competition rider. Other examples will include murals and sculptures that have been designed and made for specific areas, a made-to-measure suit or a bespoke piece of jewellery for a special occasion.

Most one-off products are, generally, hand-made using a wide variety of equipment and techniques. You will probably find this is the case with the final product you have designed and manufactured for your coursework. As a result, one-off or commissioned work is usually very expensive.

Task

Take a moment to think about the product you have chosen to manufacture for your coursework. List all the techniques you will need to employ to manufacture the item. Against this list, estimate how long it will take you to carry out each of those techniques and then total the time up.

Batch production

Batch production is the name given to the manufacture of a set number of products, ranging from just a few items to 1000s. Ceramics is an area where products are produced in batches. Producers such as Wedgwood and Royal Doulton rely on orders coming in from customers to determine what will be produced. Within a week they may produce so many thousand 10-inch plates, so many thousand 8-inch, so many 6-inch plates, and so on. These orders will be sent to the manufacturing departments, where specialist equipment is used to manufacture the items. As these products move through the manufacturing processes they will be decorated with patterns and lines on more specialised equipment, again according to the orders they have received.

For a single suit to be manufactured, a good deal of measuring is required to provide the tailor with sufficient information to create patterns for each part of the garment. The patterns will then be set out on the cloth that is then cut manually with scissors. For the manufacture of off-the-peg suits, patterns have already been established to cover a range of heights and waist sizes. Instead of a single layer of cloth being cut, multi-layers can be cut at the same time usually on a kind of band saw. This means that a number of suits can be made to the same sizes but with different coloured and patterned cloths.

Using jigs

Where a number of the same product is to be manufactured, the range of techniques used are usually (i) different, and (ii) less time consuming in the long run. For example, where a number of wrought-iron gates are required to be made all of the same size, then a jig could be used to ensure all lengths cut were of the correct size, reducing the need for additional measuring. A further jig could be set up to hold the pieces in the correct place while being joined, again avoiding the need for fiddly adjustments to be made to the set-up. The joining techniques may well be less refined, for example, a single product may well have solid riveted joints but for speedier manufacture of a number of gates then electric arc welding may be used.

Task

Go back to your list of techniques you would need to use to manufacture the one-off product for your coursework. Now consider what changes you would make to the manufacture of the product for batch production. What do you consider to be the main benefits of making these changes?

Car manufacturers produce cars in batches, again as a response to demand from customer orders placed in showrooms. As cars travel along the production line, they are sprayed the colour required by the customer and may have different internal features such as air conditioning in some, but not in all. The manufacture of cars is a highly specialised business, with equipment designed to manufacture just one kind of product; you will not find many general types of tool like spanners, drills, etc. on a production line.

Mass production (high volume production)

The standard components used in the building of a car, are, in general, produced in high volume – in some cases, very high volume. For example, for one car there may be in excess of 1000 nuts, bolts, washers and screws. Light bulbs, seat belt clips, windscreen wiper blades are all produced in high volume, otherwise known as 'mass production'.

There are very few products that are truly mass-produced. Most products are produced to order. This in turn will reduce the need for storage of over-produced products and associated costs. However, if there is one product that comes very close to mass production, it is the manufacture of polystyrene cups for vending machines. Once used they are discarded and (hopefully) recycled, and the material used again to produce more of the cups. These are produced on dedicated machines. The moulds used are very expensive, but this cost is offset by the huge quantity of cups that can be produced from them.

Continuous production

This term refers to the processes where the product is continuously being manufactured. This applies to very few processes where stopping would cause a problem; to the steel industry, for example, where the hot steel is continuously being cast into ingots for rolling into various sections.

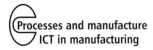
The petro-chemical industry relies on continuous refinement of crude oil to produce fuel oils, such as petrol and diesel, lubricants and materials to produce plastics.

Level of production

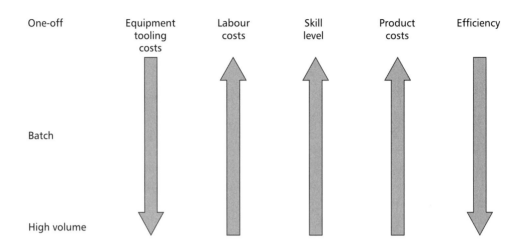

Note: Arrows show direction of increase

ICT applications in product design and manufacture

Computer-aided design (CAD)

Computer-aided design involves using computers to generate either two-dimensional line drawings or three-dimensional, photo-realistic, colour-rendered drawings.

Product designers use CAD software to speed up the designing process.

The use of CAD in pottery

Josiah Wedgwood and Sons

- Designers at Wedgwood, a pottery company based in Staffordshire, will usually create CAD drawings of pottery after creating 'mood boards' with clients and doing some initial concept sketching.
- Having decided on a particular pattern, the designer may produce a small sample of the pattern by hand, using pen, ink and paints. This pattern will then be digitally scanned so that it can be used later in CAD drawings and in the design and manufacture of decorative transfers.

The designer can then use CAD in the following ways.

- The range of standard pottery pieces used by Wedgwood, i.e. cups, plates, bowls, etc., can be drawn in both 2D and 3D and stored on the CAD system for later use. This allows designers to use these standard designs with different clients, by simply applying their pattern motif or other decoration as required. Alternatively, the designer can easily modify existing product shapes. (This can be quicker than starting from scratch.)

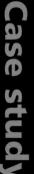

- Using editing features such as mirror, copy, rotate, move, and pattern array: patterns can be repeated and arranged around pre-drawn standard components such as vases, jugs, plates, etc. This is extremely time saving; to produce such patterns by hand could take weeks.
- Three-dimensional drawing tools enable the designer to draw bespoke designs rapidly. These can be colour rendered, have patterns and decoration applied and enhanced with lighting effects. This 3D drawing can then be tumbled so that it can be viewed from any angle. These photo-realistic images can also be placed in photo-realistic backgrounds, such as shop point-of-sale displays, table settings and so on, to give the client a realistic impression of the final design.

Rapid prototyping technology (RPT)

Three-dimensional CAD drawings can be downloaded to a machine that will make a prototype model of a design. There are many different rapid prototyping machines available today. Here are some common types.

- **Layered object modelling (LOM)** is similar to a plotter/cutter which would cut the design layer by layer in thin card or self-adhesive film. The layers are then assembled rather like a 3D jigsaw.
- **Fused deposition modelling (FDM)** is similar to a glue gun. A nozzle extrudes molten plastic and builds up the design layer by layer as it solidifies. Plastics such as ABS are often used.
- **Stereo-lithographic modelling** involves a bath of liquid resin, which uses lasers to solidify the plastic in the shape of the design.

Designers at Wedgwood will produce an FDM rapid prototype model of their designs in ABS. This normally takes approximately 24 hours. The model can then be painted to give a glazed appearance and finished with printed transfers. The end result is an extremely realistic 3D model that designers can use with clients and in planning meetings with manufacturing engineers.

Rapid prototyping greatly reduces the 'lead time' of products (the time taken from design concepts to manufacture of a product). For Wedgwood, the lead time for a six-setting dinner service can be reduced from what would have been months to a matter of a few weeks. Before the use of RPT, products would have been prototyped by making them in clay. This is a highly skilled task and would require the item to be dried, fired, glazed, re-fired, decorated with paints/transfers and fired again. This was a very lengthy and expensive process.

Virtual reality modelling (VR)

Virtual reality modelling allows designers the opportunity to see and manipulate their designs in a photo-realistic, three-dimensional environment. At Wedgwood, designers can use software that enables them to see and 'handle' products in the environment it is intended for. This can be very useful when talking to clients who can see their products in restaurant interiors, at table settings, shop displays, and so on.

At Jaguar Cars, designers and production engineers use VR systems to plan how production cells will work. The layout of work cells, the interaction of employees working on an operation together and the sequence of an assembly operation can all be planned in a virtual model. This makes tremendous cost savings for the company and, again, helps to dramatically reduce the lead-time of new models.

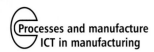

Computer integrated manufacture (CIM)

Traditionally, the design and development of a product is organised in a sequential, linear pattern. For example, a marketing department within a company passes a brief to the design department, the designers pass the designs to production engineers, who in turn pass the design to materials purchasing, and so on. If departments involved in the design and development of products work in isolation from each other, several problems can arise. For example, the designers may not interpret the original brief or product specification correctly, they may design a product that is difficult to manufacture, or the materials may be expensive and difficult to obtain. This can then ultimately result in a product that is not quite what the client had intended it to be.

Best practice in modern manufacturing work is a system known as 'concurrent manufacturing'. This is where all of the groups involved in the design and development of a product work together, right through the project.

Computers are used to assist concurrent manufacturing. Usually, those working on a project would share marketing data, specification criteria, designs and development drawings, materials' specifications and production planning over a centrally controlled database. As each member of the team works on the project, the database is updated. The use of ICT in this way leads to faster development of a product and one that meets client requirements.

CIM can also involve the central control of computer systems that organise production scheduling (the timing and sequence of production operations). This will also include the management of stock levels for raw materials and component parts and 'just in time' distribution of these around a factory.

Computer-aided engineering (CAE)

Computer-aided engineering is the use of computers to test components prior to manufacturing. Examples of this can be seen in the automotive industry, where computer models can be used to test vehicle engine or suspension parts under simulated loads. This will usually be supported with computer-controlled tests run on real components assembled in a 'test rig'.

PRODUCT ANALYSIS EXERCISE: *quality systems in manufacture*

CAD

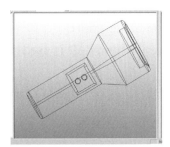

Wire frame model

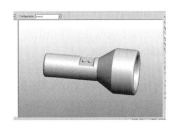

Surface model

Solid model

Study the pictures above and answer the following questions.

1. (a) Explain the purpose of wire-frame CAD drawings like the one shown.

    (b) What are the benefits of producing CAD drawings in wire frame?

2. (a) Explain the purpose of surface models.

    (b) What are the benefits of producing CAD drawings as a surface model?

3. The third CAD model is a solid model. What is the purpose of a solid model?

Exam questions

AS exam questions

1. For a material area of your choice, explain how designers might use CAD. [10]

2. For your chosen material area, explain how rapid prototyping technology is used in the development of a product. [6]

3. With reference to a product of your choice, describe how industry might use virtual reality modelling in the development of the product. [8]

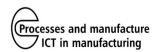
A2 exam questions

1. With reference to two different products that you are familiar with, describe how ICT is used in their design, development and manufacture. [2 × 12]

2. Designers and manufacturers use computer-aided design to reduce the time taken from design concept to production of a product.
   (a) For a product you are familiar with, describe how CAD is used in its development. [12]
   (b) Computer integrated manufacture (CIM) can greatly reduce product lead time and result in a better final product. With reference to a product you are familiar with, describe how CIM is used. [12]

Design and market influences

Introduction to Part Three

This section of AS and A2 specifications includes a broad range of areas that influence Product Design. Some of the major ones that you should be familiar with by the time you take exams include:

- design processes (how designers solve problems and communicate designs);
- communication and representation of design ideas;
- environmental issues;
- safety and quality assurance;
- human aspects of design, e.g. designing for the disabled;
- ergonomics and anthropometrics;
- advancements in technology and manufacturing processes and how these have affected product design;
- key historical designers or 'design movements' (you must be able to analyse the work of contemporary designers too).

Part Three will give you an overview of the key areas you should be familiar with. However, as in Parts One and Two, you should supplement this with further reading; some websites and books are recommended for this.

The design process

The design of products in an industrial context normally follows a process. Here is how the process works.

A need or a problem

Design activity usually arises because someone identifies a need or a problem that has to be solved. Designers often call this the 'situation' or 'design problem'. Designers can identify the problem themselves, but normally it will be presented to a designer by a client, or perhaps by marketing departments in the case of a large company.

Designers either identify a need themselves, or are presented with a need by another person, and then attempt to find a solution to meet the need.

In industry, designers may be part of a team, as they are unlikely to have expertise in all areas. In the design of a personal stereo such as the Ipod, designers would be split into teams working on styling, electronics, software and control systems, ergonomics and interface systems, and so on. These teams will have particular expertise in those areas but may also use external consultants.

Another example is in the car industry: the design of a vehicle would be shared among many designers who would be specialists in specific areas, including mechanical engineering (for engine and mechanical systems), software engineering (for engine management and instrumentation), textiles (for seating, carpets, etc.), and so on. While designers of cars would work with production engineers to plan manufacture, they wouldn't usually make the vehicles themselves.

In the commercial world, therefore, designers may work on their own or as part of a larger team. Very often, they would not be the maker of the final product.

The brief

A brief is a detailed document that sets out what is to be designed. It may contain some details of functional requirements, aesthetics, materials, safety and quality considerations, and other design constraints. The brief will guide the designers and help them formulate a specification.

The client is usually the person who has identified a need and provides the brief. This might be, for example, an individual who wants a custom-made kitchen, caravan or boat, or it may be an entrepreneur who has identified a niche for a particular product. It is the responsibility of a designer to work closely with the client to ensure that clients get the product they want.

The client, say, who wants a custom-made kitchen, is also likely to be the user. However, often the user is someone who will buy the product. It is crucial that designers make full use of market research and fully understand the needs and wants of the 'user' or 'customer base' they are designing for.

The specification

The specification is a document that expands upon the brief and sets out very detailed design constraints. It may include details of materials that the product can be made from; it may include specific functional details: what the product must do, cost limitations and references to legal and quality standards that the product must meet. The specification is referred to when designing, and in evaluating designs, to ensure that the product designed meets the original identified need.

Analysis

When designers receive a design problem and brief, they will analyse the problem looking at what they need to research in order to design successfully. Before designing can proceed, designers may have to research into unfamiliar materials or technologies, and may have to consult experts to find out whether a particular material or technology is suitable for their design project. During the analysis stage, research plans would be drawn up to detail what needs to be researched and where the information can be obtained.

Research

Research can be carried out in a variety of ways.

- Analysing similar designs to find out about useful materials, mechanisms or other features that could be incorporated into a design. This may involve testing and disassembling products.

- Testing products with consumer focus groups, to assess the problems with existing designs.

- Database searches of materials and components, to review their properties, costs, etc.

- Analysing questionnaires to obtain data on user requirements.

- Testing materials or components, e.g. testing electronic circuits to assess their suitability for a project.

Market research

Market research is a vital component in ensuring products are designed, developed and manufactured successfully. There are a number of ways in which market research is done; here are the three main ones:

- questionnaires;
- focus groups;
- field testing.

Questionnaires

Questionnaires can be used to gather a wide range of information. They can identify the need for a new product, the features of a product that consumers require, social and economic backgrounds of potential consumers, and so on. This data can be stored on computer database systems and used later to help designers draw up product specifications.

Focus groups

Manufacturers will often use groups of people that represent a cross-section of users to test and evaluate products. This may be done in a controlled environment, where the comments of the focus group can be recorded and later analysed. This type of research is particularly useful to identify potential weaknesses in designs, or to identify what features of design work well and are popular with consumers.

Field testing

Some manufacturers will test their own products against the products of rival companies to analyse their comparative strengths and weaknesses. This type of testing helps companies develop their products in order to stay ahead of competition.

Research may be carried out by designers themselves, or by specialist researchers who have particular expertise in gathering the required data.

Generation of ideas

This is the actual designing stage where sketches of ideas are drawn and evaluated. Once initial concept designs are completed, the designer will select an idea and develop it further. Some concept ideas may be drawn in a CAD system and rendered to show clients for evaluation.

Development

This is where a design idea is selected and then developed into a finished solution. This stage may include consideration of alternative ways the product could be manufactured. Component parts would be considered, such as switches, buttons, screens, and so on. The design would be refined, with attention given to the styling features as well as functional aspects. Designers may make use of scale models and mock-ups in refining the design. Prototypes may be made in order to take measurements for the manufacture of moulds or dies. Final working drawings, with details such as materials, surface finish and dimensions, will be created.

Manufacture

The maker, or to use the correct term 'manufacturer', is someone who would not design the product but would usually communicate with the designer to ensure the design is interpreted correctly, and then make the product to the correct specifications.

At the other end of the scale, the designer-makers would design the product and then make it themselves. Artisans, such as sculptors, jewellery makers, and so on, would fall into this category, because they will often design and make one-off pieces for specific clients.

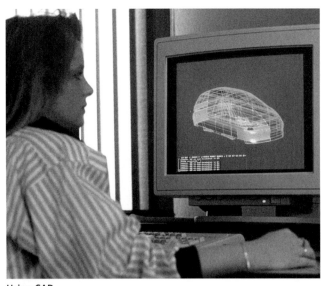

Using CAD

Manufacturing is often done by companies who are specialists in particular manufacturing technologies. For example, one company may design a package for a detergent, but the actual package may be made by a different company that specialises in plastic moulding techniques. This company may be located in a different part of the world, where labour is cheaper or raw material sources are closer.

With the use of ICT today, the maker can be geographically quite distant from the designer. The term 'global manufacturing' refers to the practice of designing a product in one part of the world and making it in another part of the world. This is often done because there may be particular expertise in an area, such as engineering and electronics in Germany and Japan, or for economic reasons because some countries have much lower labour costs than others.

Computer technology enables designs to be e-mailed to manufacturers. Video conferencing enables designers to talk to manufacturers and discuss drawings on-screen simultaneously.

Before large-scale manufacture takes place, companies usually make small production runs:

- to iron out problems during the manufacturing stage;
- to carry out product testing and evaluation.

Testing and evaluation

Prototype products, and products made for the first time, are tested and evaluated thoroughly. It is inevitable that complex products involving new technologies will have minor problems that need to be rectified before large-scale production takes place. Products need to be fully tested at this stage as, once full-scale production starts, most modern companies would want to be manufacturing to zero-defect standards.

During testing and evaluation, products are compared against the original specification to ensure that the final product meets the requirements of the specification and solves the design problem.

The design process should not be seen as a linear process with designing taking place and then evaluation. In reality, designing often takes place alongside research work; and evaluation will take place in analysing research, considering design ideas and developing a final design, in addition to evaluation of the finished solution.

Task

Think about projects that you have done and answer the following.

(a) What research methods did you use and how useful were they in gathering the information you needed?
(b) Write an example specification and explain its purpose.
(c) For a specific product you are familiar with, explain how you would test and evaluate the product.

Communication and representation of design ideas

The presentation and communication of ideas is a central part of any design course, and you should become practised at this in completing coursework. During your course, you should try to develop illustration techniques and experiment with a range of graphical communication methods.

In addition to developing the practical skills, you should ensure that you are aware of how different communication methods are used in the design and development of products.

Further reading

- *Presentation Techniques – A guide to drawing and presenting design ideas,* Dick Powell (Optima)
- *Design Graphics – Drawing and presenting your design ideas,* David Fair and Marilyn Kenny (Hodder and Stoughton)
- *Design Topics – Product Modelling,* Jennifer Cottis (Oxford University Press)

Here are some communication methods:

Mood boards

Mood boards are usually a collection of images/photos, e.g. of similar products to those being designed, colour swatches, fabric/material samples, finishes, etc. Mood boards are normally used by a designer as a style reference, when designing. They may also be used with a client, to agree on a particular style with the designer.

2D/3D sketching

These are sometimes known as 'thumbnail sketches'. They are quick, rough sketches to explore concept ideas. Designers may add critical, evaluative comments or just notes of their own thoughts as they design.

Rendering

This is the use of line, tone and often colour to make two- or three-dimensional drawings look realistic. This could be done by hand, using pencil or marker, or on three-dimensional CAD systems. Such CAD systems enable the designer to experiment with textures and colour to represent alternative materials and surface finishes.

3D CAD image

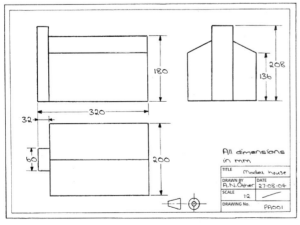

Othographic drawing

Orthographic projection

Orthographic drawings are technical line drawings, usually showing the product in a front, plan and end view. They are used to convey design details necessary for manufacture, and may:

- be drawn to a scale, e.g. 1:10;
- be dimensioned;
- include details of materials/finishes;
- be known as 'working drawings'.

Modelling

Models may be in the form of three-dimensional photo-realistic CAD models that can be used to select and develop ideas, or they may be physical models made with resistant or compliant materials.

To ensure a design is correct, designers may use scale models before it is developed any further. For example, architects would often make a scale model of a building to check proportions, aesthetic considerations, etc. They may also use the model to present designs to clients and the public.

Mock-ups

Mock-ups are a type of rough prototype, possibly made in low-cost materials such as card, MDF, plywood, etc. For example, in developing the Euro-fighter airplane, designers used a full-size plywood mock-up to test the ergonomics of the cockpit area with test pilots. On a smaller scale, product designers will often make modelling clay mock-ups for things like handles, again to test ergonomics.

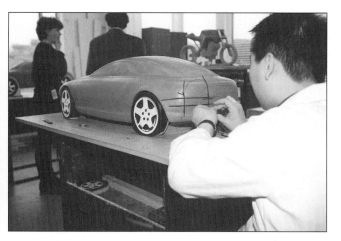

Clay modelling

Styrofoam modelling

High quality prototype model of a travel iron

Prototypes

A prototype is usually a high-quality model or functioning product that is produced to realise a design solution. This prototype would be tested and evaluated before the product is considered for production. The products you make for coursework may fall into this category, for example a concept travel iron that you may model in MDF and acrylic.

Using a presentation board

Presentation boards

Presentation boards usually display high-quality renderings, and possibly technical line drawings, showing the details of a design. A designer may use these when presenting ideas to a client, colleague or to the public. For example, architects would present their designs in the form of artist impressions and elevation drawings at exhibitions, such as public consultation events, or to local authorities to receive planning permission for projects.

Tasks

1. Make a simple 2D line drawing of a product such as a hairdryer and try rendering it with either markers or pastel chalks. (Remember to add tone to give your drawing a 3D effect.)

Tip: Your school or local library may have books on graphic presentation techniques that show you how to do this.

2. Use graphical communication books or the Internet to explain the following techniques:
   (a) sectional view;
   (b) isometric;
   (c) oblique;
   (d) perspective;
   (e) exploded view;
   (f) thick/thin line techniques.

For each technique that you find, make a sketch and some brief notes on what it is used for. Think about how you could use each of them in your coursework.

Exam questions

AS exam question

1. Explain how designers use the following communication methods in the design process:
   (a) mood boards;
   (b) isometric drawings;
   (c) 3D renderings;
   (d) scale models;
   (e) presentation boards;
   (f) working drawings/orthographic projections;
   (g) gannt charts.   [7 × 4]

A2 exam question

1. Explain in detail how **three** of the following communication methods would be used by designers:
   - colour renderings;
   - perspective drawings;
   - block models;
   - full-size mock-ups;
   - exploded views.   [8 × 3]

Design and the environment

Introduction

Manufacturers and retailers are becoming very concerned with the impact that the products they make may have on the environment. For retailers, customers have become much more aware of environmental issues, and market influences may require products to be more environmentally friendly. For manufacturers, international laws such as the Producer Responsibility Obligations 1992 (and others) set strict environmental targets for product manufacture. These can include targets to reduce packaging waste, or to recover materials from post-consumer waste (recycling materials from products once they are disposed of).

At AS level

In examinations, students often refer to plastics being used in cars. A good way of improving your marks is to make sure you refer to specific manufacturers and specific car models. It is also critical that you refer to specific components on the vehicle and show your understanding of what specific polymers are used and why. An example of the use of plastics in the manufacture of Toyota cars is given on p.161.

At A2 level

An A2 student should be able to remember specific products (at least two different ones) made from plastics and be able to assess the impact on the environment that those products might have. Such environmental impact can be assessed by the 'cradle to grave' assessment method. This involves considering how the environment would be affected by raw materials extraction, materials processing, product manufacture, distribution, use, and finally disposal – in other words, the whole life cycle of the product. An example is given on p.188.

'smart' and the environment

The smart fortwo, manufactured by DaimlerChrysler, has been designed with the environment in mind, and is a good example of a manufacturer responding in a very full way to the design and market challenges from environmental issues that impact on car manufacture and use.

The smart fortwo is an innovative two-seater, designed for city driving. It has many features that reduce its impact on the environment.

Suprex turbo engine

- This three-cylinder engine, with turbo-charger and charge cooling (forcing a proportion of exhaust back into the cylinder for combustion) gives good speed and acceleration.
- The engine is made from lightweight aluminium alloy (weighing 59 kg), and many parts of the engine are combined to reduce size and weight.

Electronic accelerator

- 'Drive by wire' technology determines the optimum position of the throttle valve related to how fast the vehicle is travelling, and how much the accelerator pedal is depressed.
- This reduces waste fuel.

Double ignition

- There are two spark plugs per cylinder to ensure as much fuel is combusted as possible.
- This results in lower pollutant emissions.

Six-speed automatic/semi-automatic gearbox

- This gives better fuel economy.
- Incorrect gear selection is reduced, as is over-revving.

Interchangeable thermoplastic body panels

- These lightweight panels reduce the overall weight of the car, reducing fuel consumption.
- They do not corrode and, as they can be changed easily for alternative colour and designs, the life cycle of the car can be extended.
- As they are thermoplastic, they can be recycled.

Interchangeable interior panels

- The interior trim can be changed if it becomes damaged or very dirty.
- This, again, can help to extend the life cycle of the car.

Upholstery

- Fillings are CFC (chlorofluorocarbon)-free and are made from recycled fabrics.
- Seat coverings and carpets use more natural materials, as opposed to synthetic.
- 95% of the vehicle's materials are recyclable.

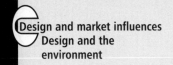

Design and market influences
Design and the
environment

Case study

Plastic parts

- Plastic parts, such as washer bottles, headlamp lenses, dash, etc. are made from thermoplastics and marked with the appropriate recycling logo.
- They are also marked with plastic identifying symbols.

Single-material components

- Engine blocks and gearbox housings are largely made from recycled aluminium.
- By reducing components that are made from combined materials, the smart fortwo is easier to recycle than many other vehicles.

Task

Use the Internet to research some of the following innovations that reduce the environmental impact of cars:

- aerodynamics (low drag);
- catalytic converters;
- bi-fuel engines (petrol and liquid petroleum gas – LPG);
- hybrid vehicles, e.g. Toyota Prius;
- water-based/solvent-free paints;
- diagnostic engine management systems.

Product life cycle

One way of assessing the environmental impact of products is known as the 'cradle to grave' approach. This assesses the whole life cycle of a product as follows.

Raw materials extraction

Considerable damage can be done to the environment by extracting raw materials. Examples of this include: de-forestation of tropical rainforests to extract hardwoods; mining for metals, which results in unsightly spoil tips and damage to the surrounding environment; oil extraction, which can result in disastrous oil spills, and so on.

Manufacturers can aim to reduce the impact of this by using recycled materials wherever possible, or materials from renewable sources such as managed forests.

Materials processing

Materials processing is the conversion of raw materials into a usable 'stock form', such as refining crude oil into plastics or smelting metal ores into ingots for later conversion, etc. This can consume vast amounts of energy and, in many cases, produce dangerous by-products and waste materials. Again, manufacturers can aim to reduce this by using recycled materials, as recycling tends to use only a fraction of the energy of that required to convert raw material. This can be seen very clearly in the example of recycling aluminium cans. The low melting point of aluminium makes it easy to recycle and much less energy is used compared to obtaining the metal from aluminium ore (bauxite).

Manufacture

Some manufacturing processes consume large amounts of energy, and many processes leave scrap or waste material. This can often be observed where pressed metal components are manufactured, e.g. aerosol cans. Huge punches and presses are required to process the sheet metal, and waste metal is produced as pieces are blanked out of the stock material. This can be reduced by careful design and cutting, and by waste recycling. Alternatively, manufacturers may seek to use processes that consume less energy, for example by using injection-moulded polymers.

Distribution

Products that are difficult to package can consume large amounts of packaging material, and can take up valuable cargo space in distribution vehicles. Manufacturers aim to make their products as easy to distribute as possible, reducing the number of vehicles required and keeping packaging to a minimum. This can be seen in drinks cans, which stack neatly together resulting in little waste space.

Use

Legislation and consumer demand is influencing manufacturers to develop products that waste as little energy as possible in their use. For example, energy-saving features on washing machines such as 40° washing cycles, half-load functions, etc.

Disposal

If products are made from recyclable materials, they can be recycled instead of being dumped in landfill sites. This can be seen in products such as the Dyson DC01 vacuum cleaner, which can be collected by Dyson and the thermoplastic parts shredded and re-used.

Some products, such as Evian mineral water bottles, are designed to be easily crushed to take up less volume in recycling containers. Alternatively, manufacturers can use biodegradable materials, such as biopol – a polymer made from natural celluloses, which break down in soil into harmless elements.

Lastly, products may be designed to have some of the parts re-used, such as single-use cameras, which will have the lens, flash and winding mechanism re-packaged with new film.

Task

Manufacturers are investigating the possibility of using shape memory alloys (a smart material) in electrical products such as mobile phones to aid recycling of components. Use the Internet to investigate and make notes on how this could be done.

Plastics and the environment

Manufacturers often prefer to use plastics rather than other materials. One major reason for this is that the use of plastics sometimes has less impact on the environment than the use of some metals or timbers.

Such environmental impact can be assessed by the 'cradle to grave' assessment method. This involves considering how the environment would be affected by raw materials extraction, materials processing, product manufacture, distribution, use, and finally disposal – in other words, the whole life cycle of the product. Two examples are set out overleaf.

Design and market influences
Design and the
environment

The Ariel washing liquid pouch

The pouch shown is made from low density polyethylene (LDPE).

Raw materials extraction

LDPE is made from crude oil. The oil is obtained from oil-producing countries that extract it from the ground or from the seabed. Surface oil wells and oilrigs are unsightly and can cause environmental problems in the future when the oil runs out and the installations have to be decommissioned as they may be contaminated with oil and by-products. More significantly, accidents when transporting crude oil can be catastrophic for the environment. This is one reason why manufacturers aim to reduce the amount of 'virgin' plastic used in a product. The Ariel washing liquid pouch has a thin wall thickness, requiring very little material, and it is possible to make such products with a high percentage of recycled LDPE obtained from post-consumer waste, such as polythene bags or other packaging.

Raw materials processing

In the case of LDPE, crude oil has to be refined into the ingredients that make the polymer. This refining process consumes a considerable amount of energy and there may also be harmful by-products made in the process. Again, if the amount of 'virgin' material used can be reduced and recycled LDPE used instead, the amount of raw materials processing required can be greatly reduced.

Product manufacture

The use of plastics to manufacture a product is of major benefit to the manufacturer, because they use little energy compared to other materials. This is because products made from polymers can be made from fewer parts, sometimes even a single piece by moulding processes. Plastics therefore usually need fewer machines to change them into a product. Thermoplastics, such as LDPE, melt at a fairly low temperature compared to other materials (such as glass, for example) and therefore require less energy to change into a molten state needed for moulding. In the case of the washing liquid pouch, the LDPE film is made by the calendering process (see p.12). This is a continuous process and can be fully automated. After the film is manufactured, the brand and consumer information are screen-printed onto the surface, the various parts cut using press knives and then seam-welded together to form the pouch.

The pouch is made from three parts: two sides incorporating a handle, and a base. The simple design and the use of only one type of polymer keep energy-use during manufacture to a minimum.

Distribution

The flexible nature of LDPE and the space left in the pouch for the contents to move enables the pouches to be packed tightly into cardboard cartons for easy distribution. Rigid bottles may have awkward shapes, leading to wastage in packaging materials and space on lorries. The pouch also takes up less space on retail shelving. In addition to these benefits, polymers are generally lightweight materials. When this is added into the distribution of millions of units, the fuel saving in distribution is considerable.

Use

A plastic product such as the washing liquid pouch is clearly not intended to have a long life cycle. It is designed as a refill for the more durable bottles made from either high density polyethylene or polypropylene. The pouch is however durable enough for its intended purpose. The gusset construction of the base allows for the package to flex, but gives it sufficient strength not to burst easily and waste the contents.

Disposal

As the consumer would throw the pouch away, the use of a minimum amount of material reduces the impact on the environment. The use of thermoplastics also enables the package to be recycled, if desired. In addition, the pouch takes up little room in refuse or in storage prior to recycling.

From this example, you can see that the use of plastics, especially thermoplastics, can be one of the best choices for the manufacturer in terms of reducing environmental impact. As manufacturers are required to meet strict environmental legislation, the use of recycled and recyclable materials is becoming more common.

Plastics in the car industry

Plastics used in the manufacture of Toyota cars

The Japanese manufacturer Toyota produces a wide range of vehicles, and in their manufacture it uses a wide range of polymers. In the Avensis and Corolla models, for example, plastics would be used for a large number of components including front and rear bumpers, door mirrors, headlamps, dashboard fascias, steering wheels, door trims, and so on.

Vehicle dashboards tend to be made from ABS (acrylonitrile butadiene styrene). This thermoplastic material is very hard wearing and more than able to cope with daily use, handling, cleaning, and so on.

Toyota Corolla

Advantages of using plastics in car manufacture
- They are available in a range of colours to suit consumer taste for vehicle interiors.
- The moulding can be given a surface texture, such as a leather-grain effect, by engraving the surface of the mould.
- The dashboards can be made in a variety of ways, the most common method being injection moulding. This enables all of the apertures for instrumentation, heater vents, controls and switches to be made at the same time in one moulding.

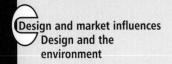
Case study: Design and the environment

- Traditionally, dashboards would have been made from pressed metals, such as steel, and either paint sprayed or covered with leather-effect vinyl or timber veneers. This is time consuming and expensive, as more machines and labour are required than for modern, injection-moulded designs.
- In volume production models, such as the Corolla and Avensis, manufacturing processes need to be as fast and streamlined as possible, therefore injection moulding is used.
- In addition, final assembly is much faster because plastics such as ABS can be joined using self-tapping screws or 'click fastenings'. This enables fast fitting of the smaller components, e.g. instrumentation into the dashboard housing.
- The use of plastics not only speeds up manufacturing but, also, the properties of plastics can be more desirable than traditional materials. For example, headlamp assemblies are usually covered in transparent polycarbonate. This has superseded glass headlamp assemblies, which had a tendency to chip or smash in small impacts.
- The use of plastics for smaller components also helps to reduce the net weight of the cars they are used in, and this will ultimately improve performance and fuel consumption.

Task

Using the 'cradle to grave' approach, choose a specific plastic car component from the list below and analyse the impact on the environment that the component may have:
- polycarbonate headlamp;
- high density polyethylene windscreen washer bottle;
- ABS dashboard fascia.

Further reading

- *Green Design, Design for the Environment*, Dorothy Mackenzie (Thames & Hudson)

- *Make The Future Work – Appropriate Technology*, the Intermediate Technology Development Group (Longman)

- **www.plasticresource.com**

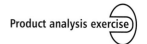

PRODUCT ANALYSIS EXERCISE: *design and the environment*

Plastics in packaging

Study the photograph of the package shown and
answer the following questions.

1. Under the following headings, explain what
   impact the package may have on the
   environment:
   (a) raw materials extraction;
   (b) materials processing;
   (c) manufacture;
   (d) distribution;
   (e) use;
   (f) disposal.

2. Explain why fast food restaurants might use paper and card as packaging materials.

3. Making reference to environmental issues, list the advantages and disadvantages of using PET
   (polyethylene terephthalate) as a material for drinks bottles.

PRODUCT ANALYSIS EXERCISE: *design and the environment*

Pepsi can

1. (a) Name a specific material or materials suitable for the manufacture of
       this product.
   (b) Explain why this material is suitable, or why these materials are
       suitable.

2. Using notes and diagrams, describe how this product is manufactured.

3. Explain how the following influence the design and manufacture of this
   product:
   (a) function;
   (b) aesthetics;
   (c) environmental matters.

Design and market influences
Design and the
environment

Exam questions

AS exam questions

1. Name a specific material used in packaging that is selected for environmental reasons. [2]

2. Explain why this packaging material is better for the environment. [4]

3. Compare and contrast a plastic product and a metal product that you are familiar with. You should make reference to the materials they are made from, their methods of manufacture and their 'impact on the environment'. [2 × 12]

A2 exam questions

1. Designers and manufacturers are increasingly concerned about the impact of their products on the environment. For **two** of the following product areas, describe a specific product you are familiar with and explain what steps have been taken to reduce their effects on the environment:
   - vehicles;
   - packaging;
   - domestic electrical products. [2 × 12]

2. Designers and manufacturers often assess the whole life cycle of their products to reduce pollution and environmental damage. For **two** different specific products that you are familiar with, assess their potential impact on the environment. You should make reference to raw materials, manufacturing methods, consumer use and disposal. [2 × 12]

3. Modern materials and technology can be used to reduce energy consumption. For **two** of the following areas, describe in detail how this has been done:
   - bottled drinks packaging;
   - domestic housing;
   - domestic kitchen appliances. [2 × 12]

Safety in product design

Introduction

Manufacturers today strive to ensure that their products are safe in order to protect the consumer. Some examples of how this is done will be outlined below.

At AS level
At AS level students need to be aware of workshop safety practices, risk assessment and Health and Safety measures taken in school workshops and common Health and Safety practices in industry.

At A2 level
A2 students should have an understanding of how safety is controlled in the manufacturing process of at least one specific product, e.g. see JCB, p.167.

Also, they should be able to describe how designers ensure safety in several products from a range of areas, for example, domestic products, vehicles, power tools, and so on.

Children's toys

Designers and manufacturers of children's toys ensure safety by ensuring:

- the toy is not finished with any toxic materials, such as paints containing lead;
- that small parts can't become detached and form a choking hazard;
- that working mechanisms, including batteries, are covered and the cover is secured with a screw fixing. This helps prevent small children from swallowing batteries;
- that there are no parts of the toy that could cause entrapment. For example, in toy prams or similar products there are safety mechanisms to prevent the toy folding accidentally and trapping fingers;
- that the toy will withstand wear and tear. For example, manufacturers of soft toys may carry out tensile tests to see if children could pull the eyes, limbs etc. off toys, creating choking hazards;
- that the toy is flame-retardant. Some soft toys may be designed from a flame-retardant material so that, for example, if a toy bear were dragged past a fire accidentally the bear would not readily ignite.

The 'CE' mark

Most toys sold in reputable stores carry 'CE' marking. This indicates that the toy meets the European Community Directive, 88/378 (Community law) for Toy Safety. Toys containing small parts would also be labelled as 'Not suitable for children under three years'.

The 'CE' mark

The British Standards Institution (BSI)

The British Standards Institution is an organisation that facilitates the processes involved and writing of UK national standards for quality and safety in products and services. BSI also represents the UK in European (CEN) and International (ISO) standards production. Companies can pay to have their products tested

The BSI Kitemark

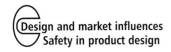
against national or international standards, and if they meet the standard requirements, and their productuion processes have been assessed and complied, they can be awarded the BSI Kitemark. The company is issued a licence to use the Kitemark on its products. This symbol of quality and safety helps to assure consumers they are buying a safe and consistent product. Companies registered with BSI have their product and their production process regularly tested.

A British Standard example

An example of a British Standard is BS EN 71. This means that the standard is both British (BS) and European (EN) and this particular standard is for toys.

BS EN 71 has eight parts as follows.

- **Part 1: Mechanical and physical properties** covers toys to ensure that they have no parts that can stab, trap, mangle or choke.

- **Part 2: Flammability** concerns toys such as Wendy houses, fancy dress costumes and soft toys. It limits the materials used to prohibit some of the more flammable ones, and ensures that if the toy does catch fire the child can drop it, or get out of it and get away.

- **Part 3: Migration of certain elements** concerns limiting the 'release' of harmful substances such as lead, cadmium or mercury from toys, if they are swallowed or chewed by a child.

- **Part 4: Experimental sets for chemistry and related activities** gives safe limits for the amount of chemicals that can be sold in such sets.

- **Part 5: Chemical toys (sets) other than those used for experiments** controls the substances and materials used in toy sets such as water-based paints, modelling clay, etc.

- **Part 6: Graphical symbol for age warning labelling** sets the standard for labelling toys unsuitable for children under three years old.

- **Part 7: Finger paints** controls the chemicals used in finger paints and minimises the risks associated with ingesting paint or prolonged skin-exposure to paint.

- **Part 8: Swings, slides and similar activity toys for indoor, family and domestic use** limits the height of such play equipment, reduces protruding parts, requires that a child or a child's clothing can't be trapped and ensures stability.

Task

Use the Internet to research the requirements of the British/European standard for playground equipment: BS EN 1176

Safety in manufacture

Manufacturers ensure the safety of employees in a variety of ways. The main methods are as follows.

Training

Employees would normally be trained in the safe operation of the equipment they use. This may include taking health and safety courses and being tested against Health and Safety Executive (HSE) standards to prove competence.

Guarding of machines

HSE regulations require most machines to be guarded to prevent employees hurting themselves. Guards range from devices to cover saw blades, such as those on band saws and circular saws, to sophisticated infra-red light beams that switch off the power to a machine if the beam is broken. These can be seen on larger machines, such as press formers or CNC (computer numerically controlled) punches.

Personal protective clothing

Depending on what tasks employees have to do, they may be issued with protective clothing, such as overalls, dust masks, safety boots, goggles and so on.

Extraction

If dust or fumes are produced in the manufacturing process, extractors are used to protect employees. Dust extraction is particularly important when machining composites such as MDF (medium density fibre), carbon-fibre reinforced plastic and glass-reinforced plastic, because the dust produced is very hazardous.

COSHH

COSHH or the Control Of Substances Hazardous to Health is a set of HSE regulations and guidance for the storage and handling of potentially dangerous materials. The regulations include details of how the materials should be labelled, safely used and stored to protect employees.

Risk assessments

A risk assessment is a calculation of the hazards associated with a manufacturing process or job, an assessment of whether the risk is a low, medium or high one, and a description of how the hazard can be controlled. All employers are legally obliged to carry out risk assessments for their operations.

Case study

> **Task**
> Carry out a risk assessment for a manufacturing process you are familiar with. (This could be either an industrial process or a school workshop-based process.)
> Two examples of safety and design in manufacture are given below.

Safety in JCB manufacture

Here is a brief summary of how safety is ensured in the manufacture of JCB earth-moving equipment at JC Bamford Ltd., in Rocester, Staffordshire.

'Safe zones'

In a busy factory such as JCB, vehicles are delivering parts 'just in time' and the parts are distributed throughout the factory using a variety of means, including fork-lift trucks. Pedestrian walkways are clearly marked throughout the factory to help minimise the risk of collision.

Safety screens

At JCB, the manufacturing process makes extensive use of laser cutting, plasma cutting and electric arc welding. Exposure to laser light can cause instant and permanent eye damage, and exposure to ultraviolet light from arc welding can cause severe eye irritation. Prolonged exposure to such UV light can also cause permanent eye damage. Heavy, black, opaque screening is used around production cells that use such processes, to protect passers-by and those working in the vicinity.

Guarding

JCB uses huge presses to fold heavy-gauge steel plate into the components used to make vehicle chassis and so on. These machines do not have any physical barriers, such as screens or cages, as they would make it difficult to

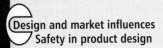
Case study: Safety in product design

load and unload the components. Instead, light beams protect them. The operator has to stand at their control pedestal, or the beams are broken and the machine won't work.

Personal protective clothing

At JCB, employees involved in arc welding have to be issued with protective welding visors. These visors darken instantly as the welding begins, and have a forced air system that blows air over the welder's face, blowing away the fumes from the welding process. This would be complemented with flame-retardant overalls, welding gloves and so on.

Use of robotics

Robots are not used to replace labour at JCB, but to take out the heavy lifting during the manufacturing process. This helps to prevent strain injury. Robots can also be used in some of the more unpleasant jobs, such as spray powder coating for the finishing of components. This removes the need to have employees working in such environments.

Job rotation

If an employee is at risk of repetitive strain injury (injury caused by doing tasks repeatedly), employees can be rotated to another production cell to work on an alternative job.

Health and safety management

JCB trains all employees to be responsible for their own safety, but also has cell managers and production managers with specific roles in the management of health and safety. They conduct detailed risk assessments for each stage of the manufacturing process, and ensure that each employee is appropriately trained. Health and safety managers also keep detailed records of any accidents and any 'near misses' that occur. (Near misses are incidents where something went wrong, e.g. a part fell off a trolley or conveyor, and could have potentially caused an accident.) These are analysed to assess how they can be avoided in the future.

Safety features in a domestic, three-pin electrical plug

A standard 240-volt, three-pin plug is packed with safety features. They are as follows.

- The wires are colour-coded: Earth (yellow with green stripe), live (brown), and neutral (blue) to help ensure plugs are wired to appliances correctly. The Earth wire is striped to aid users with colour blindness.
- Three-pin plugs are supplied with a card label fitted to the underside of the plug. This gives a clear wiring diagram to help ensure that the user correctly wires the plug up. The underside of the plug also has the letters E (Earth), L (live) and N (neutral) embossed on the surface to help the user, should the card label be lost.
- Three-pin plugs are supplied with a fitted fuse (usually 3 or 13 amp). The fuse is designed to fail if there is a sudden power surge, or if a fault develops in the product. The conducting wire within a fuse breaks and cuts the power from the appliance. This gives the user some protection from electric shock, and protects the appliance from further damage.
- The Earth pin on a three-pin plug is the longest pin. This ensures that when an appliance is plugged in, the Earth pin connects first. Should there be a fault in the appliance, the electric current will flow safely to Earth and not make the outer casing or power switches of the appliance live.
- The live and neutral pins have a small section at their base coated with a polymer sleeve. This ensures that the user cannot accidentally touch a live pin when it is being inserted or retracted from a mains socket.
- Three-pin plugs are fitted with a cable clamp. This fixes the appliance cable firmly and helps to prevent the wire at the terminals or pins becoming loose or breaking.
- The casing of a three-pin plug is usually made from urea formaldehyde. This is a hard, durable, thermosetting plastic. These properties are desirable to prevent the plug becoming damaged through general use. As a thermosetting polymer is used, the plug will retain its shape and integrity in the event of the plug becoming hot through a power surge or even a minor fire.
- 240-volt, three-pin plugs are usually marked to indicate that they meet the requirements of the British Standards Institution standard for electrical plugs and fittings BS 1363. This standard sets out a detailed specification for electrical plugs.

13-Amp three-pin domestic plug

Task

Choose one of the following products and analyse the safety features that would be incorporated in its design:
- electric jug kettle;
- electric lawn mower.

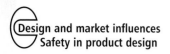
PRODUCT ANALYSIS EXERCISE: *safety in product design*

1. For each of the following features on a family car, describe how they improve safety. Use notes and diagrams under each heading.

   (a) anti-lock brake system (ABS);

   (b) laminated glass windscreen;

   (c) adjustable headrests;

   (d) side impact bars;

   (e) air bags.

Exam questions

AS exam question

1. (a) Describe the safety requirements of a soft toy suitable for children under three years old.　　[5]

   (b) Explain how manufacturers would test soft toys to ensure that they are safe for young children to use.　　[15]

A2 exam question

1. (a) With reference to the manufacture of a specific product that you are familiar with, explain in detail how manufacturers ensure the safety of employees.　　[12]

   (b) Produce a risk assessment for **one** of the following processes:

   • wood turning.

   • electric arc welding.

   • using adhesives.　　[12]

Design for the disabled

Introduction

In an inclusive society, it is extremely important that designers take into account the different needs of different groups of people. The disabled make up one such group of consumers – though it's important to note that, of course, different disabled people have different needs.

Adapted products

Many domestic products have been adapted to meet the requirements of the disabled and, as companies realise the market demand, more and more products are being especially designed with the disabled user in mind.
Some examples of adapted products include:

- kettles with tipping devices for one-handed operation (to assist amputees or stroke victims);
- tap lever attachments to aid opening and closing taps (especially helpful for arthritis sufferers or stroke victims who may find it difficult to grip);
- hand controls – retrofitted to cars to enable acceleration and braking instead of using foot pedal controls.

Some products specifically designed for the disabled might include:

- stair and bath lifts;
- lightweight and manoeuvrable sports wheelchairs;
- clothing using Velcro fastenings (much easier for arthritis sufferers to use than buttons).

> **Task**
>
> Use the Internet to search for products that have been designed for the disabled. Sketch or paste images of them, making notes on the features that may help the disabled.

Public buildings

Legislation now requires architects involved in planning new public buildings, such as schools, libraries and hospitals, to ensure access and facilities for the disabled.

Activity: assess your school

Use the following checklist to assess how well your school has been designed to meet the needs of the disabled.

Car parking

- Are there clearly marked disabled parking spaces near the entrances?
- Are the car parking spaces wide enough apart to allow for wheelchair users to get in and out of cars?

Entrance

- Is the car park on the same level as the entrance? If not, has a ramp been installed to help wheelchair users?
- If there is a ramp, how steep is the slope?
- Are there steps to negotiate entering the building? If so, has a handrail been installed to help people who are unsteady on their feet?
- Is the door easy to open? Is it automatic?
- Is there a threshold or step in the doorway? Could this be difficult for wheelchair users to get over?

Corridors

- Are corridors wide enough to allow for wheelchair users and able-bodied users to pass comfortably?
- Are there internal doors? If so, are they easy to open and are they held back on catches or magnets?
- Are there refuge points for wheelchair users in the event of a fire?

Toilets

- Are disabled toilets provided? If so, are they clearly signposted and positioned in an accessible place?

If a disabled toilet is available, note how it may have been adapted.

- If a door is fitted, is it a lightweight sliding door and wide enough to allow wheelchair access?
- Look at the level of sinks (they should be at wheelchair height).
- Taps should have a lever attachment on the top to make it easier to turn them on and off.
- Toilet cubicles should be wide enough to take wheelchairs and allow the user to slide from the wheelchair onto the toilet. There should be grab rails near the toilet to help with this.

Lifts

If lifts are provided, note how they may be designed to help the disabled user.

- Is the call button at a sensible height for wheelchair users?
- Is the lift door wide enough for wheelchair users? Does it close slowly and re-open if the door is obstructed?
- Is there a threshold or step between the floor level and the interior floor of the lift?
- Are the interior call buttons easily accessed? Are they illuminated to indicate to the deaf that the floor has been selected? Are they engraved in Braille for the visually impaired?
- Are there audible indicators for the door closing/opening and for the different floor levels?
- Are there handrails fitted to the interior of the lift to provide a steady movement?

Task

Analyse a public telephone and note how it may have been designed to accommodate disabled users.

Occupational therapy workshops

Occupational therapy (OT) departments within hospitals help people recover from an illness, or come to terms with a permanent disability.

Most hospitals have an occupational therapy department that helps many patients in the recovery from strokes, which cause long-term paralyses or partial weakness down one side of the body. This can mean stroke victims may have very poor hand-eye co-ordination or manipulation skills in one hand.

At the hospital, the OT department has small workshops where the patients make simple products to help them in the home. These may be devices to open taps, grip and pour bottles, pour kettles, peel and slice vegetables, slice bread, and so on. In making the living aids, the patients are learning how to use tools and equipment with only one hand functioning normally. This in itself is good therapy and, at the same time, they are making products to help themselves. In the same workshops, patients also make simple games such as solitaire and draughts and then play them. Activities like these help patients to recover and practise manipulation skills.

Occupational therapy in a kitchen environment

The Baygen Freeplay Radio

The Freeplay radio, invented by Englishman Trevor Bayliss, is a clockwork-powered radio that requires no batteries. Bayliss, watching a programme about AIDS in South Africa, saw that Africans were not hearing advice about prevention of AIDS because, while most people own radios, batteries are very expensive for ordinary people to purchase.

The initial prototype of the radio was further developed by engineers at Brunel University, in order to make the radio play for longer and louder. After trialling prototypes in South Africa, further development and subsequent manufacture of the radio was sponsored by the charity Liberty Life that helps disabled people in South Africa. The radio is manufactured in a factory in Capetown, South Africa, by a workforce that is largely made up of disabled workers. The radio has relatively few parts and can be assembled by workers who are blind, or have other disabilities that might make employment elsewhere difficult.

The Freeplay radio is a successful product, sold not only as a functional product in South Africa, but also a desirable consumer product in other international markets. The quirky application of an old technology to a modern product makes the radio stand out from others on the market.

Further reading

- The Disabled Living Foundation is a UK charity with comprehensive information about disabilities, products for the disabled and ways in which homes, gardens, cars, and so on can be adapted for the disabled. Its web address is: **www.dlf.org.uk**

- *The Measure of Man – Human Factors in Design*, Henry Dreyfuss Associates (Whitney Library of Design)

- *Design Topics – Human Factors*, Steve Garner (Oxford University Press)

PRODUCT ANALYSIS EXERCISE: *design for the disabled*

Study the photograph of the kettle tipper. This is a device used to help disabled people tip a kettle in order to pour drinks safely.

1. Who would find this kettle pourer a useful item (what kind of disabilities)?

2. Use notes and sketches to show how the kettle tipper functions.

3. Use the Internet to research living aids for the disabled that would help with the following routine activities:

- turning water taps (with arthritic hands);
- dressing (with the use of only one hand);
- making a cup of tea (with poor eyesight);
- driving a car (without the use of legs).

Make notes and diagrams to show how each living aid functions.

Exam questions

AS exam question

1. (a) List the modifications that might be made to a public building, such as a school or leisure centre, to meet the needs of disabled users. [10]

(b) Describe how a public telephone should be designed to meet the needs of disabled users. [10]

A2 exam question

1. (a) Use notes and diagrams to explain in detail how the design of kitchens can be developed to meet the needs of the disabled. [12]

(b) Use notes and diagrams to show how the controls of a car can be adapted to meet the needs of disabled drivers. [12]

Ergonomics and anthropometrics

Ergonomics

Ergonomics concerns the interaction between the human body and products, systems or environments. Product designers are particularly concerned with making products that are easy to use.

Ergonomists are designers specialising in ergonomics; they may design:

- **systems:** for example, the layout of a restaurant kitchen or a manufacturing production cell, in order to make them more efficient and reduce strain injury risk;
- **environments:** for example, the interior of a car or aircraft, to ensure comfort of the user and an efficient interface between the control systems and the driver or pilot.

Anthropometrics

Anthropometrics involves using body sizes to improve the ergonomics in products, systems or environments. For example, designers of items such as personal stereos would look at data for hand sizes in order to ensure that the product can be comfortably held, and that the controls are positioned to allow easy operation with finger tips.

Anthropometrical data taken from the measurements of hundreds of volunteers is normally recorded as percentiles. The average size is known as the 50th percentile. Most design activity is for the body sizes between the 5th and 95th percentile, which would take in the majority of the population.

Ergonomics and anthropometrics involve a wide range of areas. Some of these can be illustrated if we analyse some familiar products.

PRODUCT ANALYSIS EXERCISE: *ergonomics*

Mobile phone

Study a mobile phone and note the following ergonomic features:

1. How comfortable is it to hold? Does it fit into the hand neatly? Has it been shaped to fit into the hand?

2. Are the number/letter keys spaced sufficiently to allow easy dialling?

3. Are the keys laid out in a logical sequence? (This can help the user to make calls/texts and prevent errors.)

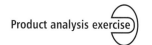

4. Are the keys backlit to enable the user to dial in dark conditions?

5. Is the screen backlit?

6. How is colour used on the keys (usually green to answer a call, red to end a call – familiar colours for 'go' and 'stop')?

7. Are simple ideograms or pictures used on the buttons or menu to help the user recognise features, for example power button, end call, volume, battery level?

8. How clear are characters/digits on both the keys and the screen?

9. How easy is it to navigate around the menu? Is this laid out in a logical order?

PRODUCT ANALYSIS EXERCISE: *ergonomics*

Car interior

Study a car interior and note the following ergonomic features.

1. Adjustable seats – to take account of different body sizes and driving positions. How easy are they to adjust? What parts are adjustable and what is the range of adjustment?

2. Shaped and padded seats to provide comfortable seating.

3. Adjustable mirrors.

4. Layout of the dashboard. Note that the instruments such as speedometer, temperature gauge, and so forth, are usually visible either over or through the steering wheel. Controls such as heater controls, light switch, hazard warning light switch, and so on, are within easy reach.

5. Use of colour on instruments. Note the use of red and blue on temperature controls. Red is used on engine warning lights, for example for the low oil indicator and brake fluid level indicator. Red is used on the speedometer to indicate the 30 m.p.h. mark and on rev counters to indicate high level engine revs.

6. Light level of instruments. The brightness of instrument lighting can usually be adjusted to allow for different consumer preferences.

7. Heating/air conditioning. Interior temperature control is an important aspect of maintaining passenger comfort.

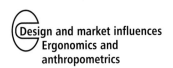

PRODUCT ANALYSIS EXERCISE: *ergonomics and anthropometrics*

Computer workstation

1. Sketch a computer workstation and label the features A to R, listed below.

2. Under the sketch, make brief notes to explain the ergonomic and anthropometrical features identified by each label.

Desk and computer
A Foot rest
B Leg room under table
C Computer grade lighting
D Ambient temperature, + 15°C
E Adjustable monitor
F Wrist rest
G Screen filter
H Document holder
I Window blinds

Chair
J Cushioned seat and backrest
K Adjustable height/gas strut
L Adjustable backrest
M Seat avoids pinching knee joint
N Backrest allows free movement of shoulders
O Fabric made of breathable materials
P Five spoke wheels
Q Backrest hugs curvature of spine

PRODUCT ANALYSIS EXERCISE: *ergonomics and anthropometrics*

Personal stereo

Sketch a personal stereo and headphones like the ones illustrated overleaf. Label and make notes on the following features:

Personal stereo

Overall dimensions relate to average hand size.

Rounded corners for comfort.

Lightweight materials.

Ideograms used on fast forward, and so forth.

Textured wheel for volume adjustment.

Raised buttons for radio/stereo selection.

Headphones

Flexible steel band adjusts the headphones to fit different head sizes.

Wire long enough to allow freedom of movement.

Foam padding used on earpiece.

Earpieces swivel to 'best fit'.

Earpieces can be adjusted vertically.

Personal stereo

Enlarged play button.

Easy battery access.

Volume restricted to prevent ear damage.

Fingernail recess for tape removal.

Pocket size.

Miniature earpiece type

Sculpted to fit average.

Stem fits in natural slot of ear lobe.

May have foam covering for comfort.

Flex has one piece longer to go around back.

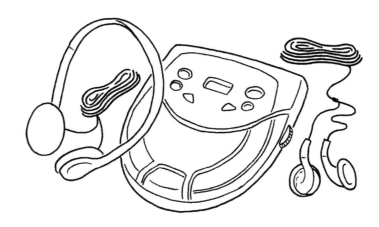

Further reading

- *The Measure of Man – Human Factors in Design*, Henry Dreyfuss Associates (Whitney Library of Design)

- *Design Topics – Human Factors*, Steve Garner (Oxford University Press)

Exam questions

AS exam questions

1. Define what is meant by the terms 'ergonomics' and 'anthropometrics'. [5]

2. Describe the anthropometrical requirements of a jug kettle. [5]

3. Using a labelled diagram, describe the ergonomic features of a jug kettle. [10]

A2 exam question

1. With reference to specific products from the product areas listed below, explain in detail the ergonomic and anthropometrical requirements of each product:

   (a) bicycles;

   (b) steam irons. [2 × 12]

Major developments in technology

Introduction

Woods and metals have been used in the manufacture of products for hundreds of years. Yet it is only the past 100 years that has seen significant changes in materials and subsequent manufacturing methods. One of the biggest changes to the materials available is the introduction and development of plastics – these have been available commercially for the past 70 years.

The introduction of plastics has seen not only the development of a huge range of polymers (each with a particular set of materials properties) but also a range of manufacturing processes tailored to the processing needs of the new materials. These processes include compression moulding, injection moulding, blow moulding, vacuum forming and thermoforming, etc.

Over the last 70 years there has also been significant development in electrical and electronic products, which has worked hand-in-hand with the materials and processing developments to produce the commercial products we have today.

Three examples of products to illustrate these major developments in technology are shown below: radios, batteries and TVs.

Technological developments and the radio

1930s

Radios became commercially available from the late 1920s and early 1930s. At that time, there was still a strong reliance on craftsman-made furniture and the radio casing reflected that. The electronics inside consisted of a mains

1930s radio

(240 volt AC) powered circuit using thermionic valves. These were the forerunner of the transistor, and operated by emitting electrons when the valve components were heated. An internal grid controlled the flow of electrons (and therefore the current), making them an effective amplification stage for the incoming sound signal.

A number of amplification stages were used, together with a local oscillator, an aerial, and a big (approx. 200 mm) diameter loudspeaker. Other components included dozens of resistors and capacitors, and a large amount of copper wire used for connecting all of the components together.

Internal construction showing valves

There was a large space around the components to allow for the dissipation of the heat (40 watts or so) generated by the valves and other components. The whole product was extremely heavy due to the weight of the wooden case, the metal chassis supporting the electronic components (including the valve holders) and the large transformer used to alter the voltage from 240 volt to 6.5 volt for heating the valves.

This radio would have been very expensive to buy, probably costing the equivalent of a couple of months' wages, although by today's standards the quality of reception would be considered poor.

1940s–1950s

During the 1940s–1950s, radios tended to be made from the newly developed plastics. In particular Bakelite (a thermosetting polymer) was being used for various external components, such as volume- and station-control knobs.

The transistor

Around about the same time, in the early 1950s, the transistor was introduced. These are made from two kinds of semiconducting materials, i.e. materials that will only conduct electrons when a particular voltage is applied. These two materials are known as n-type and p-type materials, and refer to the special arrangement of particles in either silicon or germanium. Transistors work as an electronic switch: switching on when the current at the p-n-p junction is above a minimum value. They can be used to control the flow of electrons, making them effective at providing an amplification stage for the radio.

The introduction of transistors revolutionised the electronics industry. Since the transistor needed considerably less power than a thermionic valve, their use allowed circuits to be developed that required considerably less power. This also resulted in the reduction in the size of supporting components, such as resistors and capacitors. The radios could now be powered by a 9 volt battery, and so became truly portable.

This miniaturisation of electronic components required a new form of circuit construction. The heavy metal chassis and solid copper wires were now being replaced by a pattern of conductors fixed to one side of a substrate – very similar in structure to fibreglass. Each electronic component needed to be placed and soldered by hand, although some level of mechanisation existed for creating the printed circuit boards.

1960s

Up until, and including, the 1960s the cases of radios were still being made from a wood material in the form of bent wood veneers. These were a much coarser construction than the cabinets of the 1930s, and were consequently covered in a plastic 'leatherette' (mock-leather) material.

The sound quality of these radios was much improved, with sound being more suited to listening to music.

1960s radio

1970s

From the 1970s onwards there was increased use of thermoplastics such as acrylics, and acrylic-based polymers such as ABS (acrylonitrile butadiene styrene). With the introduction of these plastics, it was possible to use injection moulding to manufacture the cases of the radios. These modern plastics have replaced the Bakelite and laminated/plywood cases of the past. They have also made it possible to mould-in styling and ergonomic features, while the continued reduction in size and power consumption of the electronics has enabled a much-reduced size of the product.

The method of manufacturing with plastics has provided huge benefits, as they can include mountings for circuit boards, handles, fixing brackets and strengthening ribs. Also, this can be done in an all-in-one single moulding process.

A further benefit of manufacturing with plastics is the little finishing required, along with the ability for the product to be self-coloured – achieved by mixing pigments with the plastic granules prior to manufacture.

Electronics have become much smaller, even miniaturised, and will fit into a much smaller space. Hundreds, if not thousands, of transistors and other supporting components can be grown onto a single silicon chip – called an integrated circuit (IC). The energy required for these components to function is extremely low, thereby reducing the need for a large battery to power the unit or

Design and market influences
Major developments in
technology

for extra space to dissipate any excess heat. The product has now become much smaller, fitting into the hand or pocket.

1970s – recent years

The introduction of the microchip in the 1970s revolutionised the design and manufacture of electronic products. The microchip has made it possible to miniaturise products like radios and to increase the range of features available to the consumer. The Sony Walkman exemplifies this.

The modern Walkman is truly pocket-sized, brimming with ergonomic features:

- very comfortable to hold;
- lightweight;
- simple to use.

The carbon rod aerial used in the 1960s radio has been replaced by using the wires to the headphones as an aerial, thereby reducing size and weight of the final unit further.

Many electrical goods now have a GUI (graphical user interface), or a visual display either in LED (light emitting diode) or LCD (liquid crystal display), made possible by the microchip. We take it for granted that the quality of sound can be adjusted through a graphical equaliser to suit the listener's taste. Also, in CD variants of the Walkman, there is the opportunity to play tracks in random order, intro scans, etc. All of these programmable features are made possible by the microchip.

In recent years there has been an explosion in the development of digital sound recording and playback equipment, due to advances in microelectronics, for example Minidisc, MP3 and memory sticks.

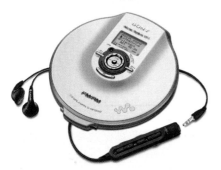

A 1970s Sony Walkman and a
1990s Sony CD Walkman

Technological developments and the battery

The reduction in energy requirements has made a whole range of electronic devices portable. Whether they use standard alkaline batteries or rechargeable nickel-cadmium batteries depends entirely on the application.

The reduction in size of portable equipment such as mobile telephones has been helped by developments in battery technology and their subsequent reduction in size.

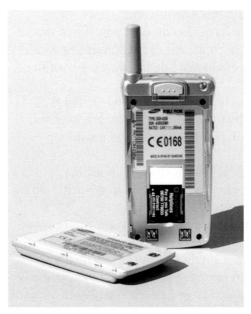

How do batteries work?

Batteries are made up essentially of two electrodes in an electrolyte. Energy is produced by one of the electrodes decomposing into the electrolyte.

Electrolytes can be acids, alkalines or salts. When salts dissolve they release positive and negative ions, which are then free to carry electrical charge between electrodes. Ions are electrically charged atoms.

One of the electrodes will then collect atoms, while the other electrode will lose atoms and slowly decompose into the electrolyte. The rate at which it decomposes depends on the materials used for the electrodes and electrolyte.

How a battery works

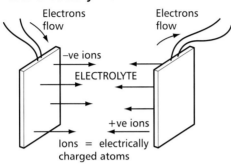

Types of battery

Modern types of battery use a variety of chemicals to provide energy, as shown in Table 18.

Table 18: Types of battery

Battery type	Made up of	Uses
Zinc–carbon	Zinc and carbon electrodes with acidic paste electrolyte	Inexpensive AA, C, D sized dry cell batteries
Alkaline	Zinc and manganese oxide with alkaline electrolyte	More expensive brands of batteries, e.g. Duracell and Energiser
Lithium photo	Lithium, lithium iodide and lead iodide	Used in cameras, copes well with power surges for the flash
Lead–acid (rechargeable)	Lead and lead oxide electrodes with acidic electrolyte	Used in cars, etc.
Nickel–cadmium (rechargeable)	Nickel hydroxide and cadmium electrodes; potassium hydroxide is the electrolyte	Used in a variety of products as rechargeable standard-sized batteries
Nickel–metal–hydride (NiMH) (rechargeable)	Alloy and nickel-hydroxy-oxide electrodes; potassium hydroxide electrolyte	Generally replacing nickel–cadmium batteries
Lithium ion (rechargeable)	Lithium ion and lithium atom electrodes. The electrolyte also contains lithium	High-end laptops and mobile telephones; good power to weight ratio
Zinc–air (rechargeable)	Zinc and oxygen (from air) are the electrodes. Electrolyte in potassium hydroxide	Lightweight and rechargeable
Zinc–mercury oxide	Zinc and mercury oxide electrodes. Electrolyte in potassium hydroxide	Small sized; used in hearing aids
Silver–zinc	Silver/manganese oxide and zinc electrodes. Potassium hydroxide in the electrolyte	Aeronautical applications; good power to weight ratio
Metal chloride	Typically electrodes are sodium and sulphur with ceramic alumina as the electrolyte	Used in electric vehicles

Technological developments and the TV

Analogue television

Televisions have been around for well over half a century. The most common type of TV remains the analogue type. A composite signal is fed to the television from a transmitting station or from a videocassette recorder. The signal being transmitted comprises a beam of varying intensity as it creates a 'line' providing:

- the image on the screen;
- a signal to send the beam back to the beginning of the line;
- a signal to send the beam back to the top of the screen to start the cycle again.

The signal also includes sound information.

The signal for colour televisions includes the additional information that is used to switch the phosphor dots on or off to provide the image. The signals received by the colour television can be received through an aerial, VCR or DVD player, cable television or satellite via a set-top decoder (through the aerial socket at the back of the TV).

Digital television

The signal received by a digital television is in the form of 0s and 1s – in the same way that computers receive their signals. Being digital, the image produced is a lot more stable and of very high resolution. The signal conditioning electronics are more sophisticated in a digital TV because of these higher resolutions.

HDTV (high definition television)

HDTV provides a higher resolution, producing greater clarity along with high-quality sound.
Types of HDTV include the use of:

- thin film transistor (TFT) technology;
- LCD technology;
- plasma technology.

With these types of television the depth of the unit is much reduced, therefore reducing the weight of the unit resulting in flat screens that can be supported on a wall just like a picture.

Projection television

TFT and plasma technology can provide large screen sizes. However, where screens are larger than 40 inches projection television is used.

A small CRT (cathoderay tube) or liquid crystal display forms an image that is shone onto a large screen by one of two methods:

- rear projection;
- front projection.

In rear, or reflective, projection the image is projected onto a reflective surface and then onto a screen. All components are housed in one unit, for use in the home as a home theatre system.

Front projection, or transmittive, TVs require the projector to be used in one part of the room with the screen at the other, much in the same way as digital PC projectors are used.

14″ digital television (top) and a wall-mounted plasma screen

Rear (reflective) projection

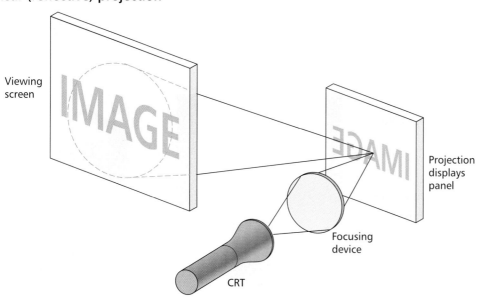

Viewing screen

Projection displays panel

Focusing device

CRT

Front (transmittive) projection

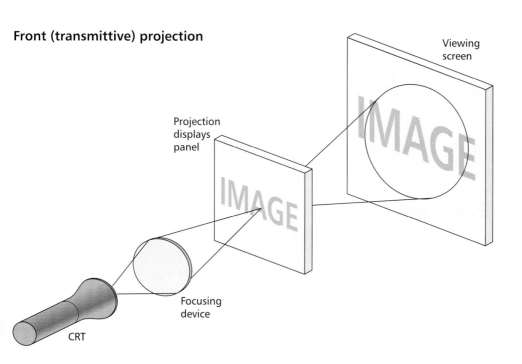

Viewing screen

Projection displays panel

Focusing device

CRT

PRODUCT ANALYSIS EXERCISE: *major developments in technology*

Mobile phones

Study the photographs that show a 1980s 'brick' mobile phone and a contemporary mobile phone.

1. Make notes under the following headings, explaining what advancements have been made in technology to enhance the performance and function of mobile phones.
   (a) Batteries.
   (b) Visual displays.
   (c) Microelectronics.

2. Explain how developments in technology have led to improvements in the ergonomics of mobile phones.

3. Explain how developments in smart materials can make electrical products easier to recycle.

Exam questions

AS exam question

1. With reference to the following developments in technology and example products, explain how the product's function and performance have been improved: [4 x 6]

Development	Product
(a) Surface mount electronic components	Personal stereos
(b) Thermochromic pigments	Thermometers
(c) Colour liquid crystal displays	Mobile phones
(d) Fibre-reinforced polymers	Safety clothing, e.g. workshop gloves

A2 exam question

1. The performance and function of products can be enhanced by changes in materials and technology. With reference to **two** specific products that you are familiar with, from the product areas listed below, explain what developments in materials and manufacturing processes have taken place and outline how these have improved the performance and function of the products;
   - consumer electrical goods;
   - power tools;
   - safety clothing.   [2 x 12]

Product life cycles and historical influences

The life cycle of a product can be divided into several stages. These are:

- introduction;
- growth;
- maturity;
- decline.

How long a product's life cycle lasts for depends on two important factors:

1. changes in materials and technology (often known as the 'technology push').

2. changes in consumer demand (often known as 'demand pull').

- **Introduction:** the introduction stage of a product's life cycle is the period when a product is newly released onto the market. At first sales can be slow, as consumers may not recognise the benefits of a new product. This was seen in the early 1980s with the introduction of the Sony Walkman. At the time, large 'ghetto-blaster' stereos were fashionable and it was some time before the Walkman began to sell in large quantities.

- **Growth:** as advertising takes effect and consumers see the benefits of new technology, sales start to rise. During this stage of a product's life cycle, competitors may start to introduce their brand of the product. Again, the personal stereo market illustrates this with most of the major electronics companies having their own version of the original Sony classic design.

- **Maturity:** at this point in the life cycle, sales begin to level off. The market becomes saturated with competitor designs that may have different or improved features. Major companies, such as Sony, monitor the market very carefully and have new designs ready – in order to maintain their market share.

- **Decline:** as new technology develops, products can be made obsolete. This is clearly illustrated with the replacement of vinyl records and audiocassette tapes with the compact disc. More recently, Sony developed the minidisc, reducing the size of personal stereos. Then they developed the MP3 player, making a personal stereo truly pocket size. The introduction of Internet marketing of music and the MP3 player has the potential to make CDs obsolete. 'Consumer pull' is clearly contributing to this as a whole new generation routinely downloads music from the Internet, rather than purchase CDs.

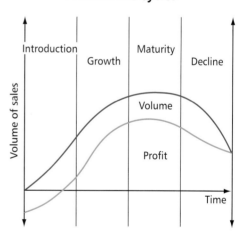

Product life cycles

Planned obsolescence

Cynics argue that many manufacturers build obsolescence into products so that consumers will have to upgrade on a regular basis, and thus maintain profits for the company. Today, many electronic products, such as mobile phones and personal computers, do seem to have a limited life; this is, perhaps, due to rapid advancements in microelectronics and communications technology. Such advancements do tend to result in products being rapidly superseded by other products offering enhanced features or faster operation.

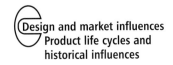
Design and market influences
Product life cycles and
historical influences

Some products need to have a built-in obsolescence for safety or hygiene reasons, for example, the hypodermic syringe or disposable razor. Others, such as cars, may have a shorter life cycle than is actually possible, in order to keep the overall cost of the vehicle at an affordable level. It is possible, of course, to make cars from very durable materials, like stainless steel, but it is very expensive.

The influence of fashion

Consumer fashions, trends and the demand to keep up with the latest technology, all help to contribute to obsolescence. Many people change their wardrobe on an annual basis, and their home interiors every 3 to 5 years, and this therefore feeds the demand for new products.

The influence of fashion and trends on product design cannot be over-estimated. Manufacturers today employ agencies to predict what the latest fashions will be. Such agencies will advise on colour and fabric trends for interiors and for the fashion industry. This results in seasonal colours and fabrics being manufactured together with a full range of coordinating accessories. As consumer tastes change, the season's colour can become obsolete.

Fashion can be influenced by changes in society and the economy. During the 1980s, the UK economy was booming and there was a rapid increase in demand for 'designer' products to furnish smart city apartments of upwardly mobile consumers. It was fashionable at the time to have interiors that reflected a business-like lifestyle, so furniture (often black leather and chrome, clean-lined, and simple) would be placed in rather clinical white-painted rooms, resembling offices. Today, people generally have more individual tastes and demand products that will enable them to achieve an individual look. This poses a challenge to manufacturers, who have to respond quickly to market pressures.

Historical influences on product design

You need to be familiar with influential designers or design 'movements' that have made a significant contribution to product design. It is only possible within this book to introduce a small number. You should also research design history in other textbooks or on the Internet.

Further reading

- *Icons of Design – The 20th Century*, edited by Volker Albus, Reyer Kras and Jonathan Woodman (Prestel)
- *A Century Of Design – Pioneers of the 20th Century*, Penny Sparke (Mitchell Beazley)
- *Bauhaus-Bauhaus archiv*, Magdalena Droste (Taschen)
- *Marcel Breuer*, Magdalena Droste and Manfred Ludewig (Taschen)
- *Bauhaus*, Judith Carmel-Arthur (Carlton)
- *Memphis*, Brigitte Fitoussi (Thames & Hudson)
- *Philippe Starke*, Judith Carmel-Arthur (Carlton)

Task

Advancements in electronics and communications technology have led to improved consumer products. Using a specific product you are familiar with, describe the technological developments that have taken place and the benefits they have brought to the consumer. You should refer to a specific product from one of the following areas:

- mobile phones;
- personal computers;
- home entertainment.

The Bauhaus and Modernism

The Bauhaus or 'Building House' was a school of art and design founded in Germany in the 1920s by architect Walter Gropius. At the Bauhaus, students followed a foundation course where they experimented with materials, form and colour (especially new materials for a machine age, geometric forms and primary colours). Students then specialised in areas such as architecture, furniture, textiles, graphics, metalwork and so on, working with leading experts in those fields such as Marcel Breuer, Wassily Kandinsky, and Ludwig Mies van der Rohe.

The work of the Bauhaus was very much influenced by a set of design principles.

- **'Form follows function'**: an object's appearance should be influenced mainly by what it is intended to do. In other words, a product's appearance should not be the most important factor. Above all it should function well.

- **'Everyday objects for everyday people'**: products should be affordable to a wide range of consumers.

- **'Products for a machine age'**: products should be designed to be made with the use of mechanised processes and modern materials.

- **Geometrically pure forms**: designs should use vertical, horizontal, geometric shapes and clean lines with no fuss or clutter. They should also use basic tones and primary colours.

The Bauhaus tutors and students went on to design what have, in many cases, become design classics such as the 'Wassily Chair' (designed by Marcel Breuer in 1924) and the Barcelona Chair (designed by Mies van der Rohe for the Royal Pavilion at the 1926 Barcelona Exhibition). These chairs, and many other Bauhaus products, are still in production today.

The Bauhaus moved to Chicago in the 1930s to flee Nazi persecution. Many of the Bauhaus designers went on to become very influential in shaping American architecture. Mies van der Rohe was one of the pioneers in using reinforced concrete and glass to make affordable, open-plan buildings. This style of architecture became the principal method of constructing high-rise buildings.

In addition to these lasting designs, the most influential contribution to design that the Bauhaus made was the principle of 'form follows function'. The idea that products can be made ergonomically correct, using appropriate materials and with the minimum of applied decoration became the doctrine of what is known as the 'Modern Movement' or 'Modernism'. Designers whose work reflects this style were known as Modernists. The principles of good design form the basis of contemporary industrial practice.

Marcel Breuer's B32 chair (left) and the Barcelona Chair by Mies van der Rohe (right)

Post Modernism

Post Modernism is a 'style label' associated with groups of designers, architects and artists whose work reacts against the principles of the Modern Movement. The basic principles of Post Modernism include:

- focus on aesthetics rather than the function of a product;
- use ornamental and decorative finishes to enhance aesthetics;
- design to appeal to fashion, popular consumerism and youth culture;
- borrow and mix styles from other periods, such as ancient Egyptian;
- draw on influences from the media and fashion, and use everyday materials.

The Memphis Group

Carlton room divider in plastic laminate by Ettoire Sottsass. 1981 Collection "Memphis Milano"

The Memphis Group was a group of designers based in Milan, Italy during the 1980s. Its founder member, Ettoire Sottsass, produced several designs that typify Post Modernism.

The Carlton Dresser, designed by Sottsass in 1981 and shown here, is typical of the Memphis Group and of the Post Modernist style. Its design is intended to give maximum visual impact rather than to function as a dresser (used to store and display items). This is achieved by the use of a striking mixture of bright colours and the angular structure that is reminiscent of a piece of Ancient Egyptian.

The angular structure of the shelves makes the dresser barely functional. There is little usable space on the shelves on which to display anything, and there are only two small drawers to store things in. The dresser uses everyday materials such as MDF, finished with melamine laminates to provide colour, and the 'stone effect' base.

The Carlton Dresser is intended as a statement piece. It looks, like many Memphis products, at home in a museum of art and design rather than in a home as a functional piece of furniture.

The Etruscan chair pictured right, was designed by Danny Lane in 1984. Lane is an established designer whose most recent works include a glass balustrade at the Victoria and Albert Museum in

London, and a stone, steel and glass sculpture at the GlaxoSmithKline headquarters. Lane specialises in using industrial glass and contradicting materials and forms.

The Etruscan chair is also a typical Post Modernist piece. Although it would have some function as a chair, it is really a work of art. It is an aesthetic piece, designed as a one-off gallery exhibit, or to be made in small numbers for private clients. The Post Modernist aspects of the chair can be shown if it is briefly analysed.

- The use of glass is normally associated with quite a different function – glass is thought to be brittle and sharp.
- The transparent glass allows you to see the sculptural qualities of the legs, but it is a cold, hard surface to sit on.
- The use of glass with a serrated edge gives an interesting aesthetic feature but, again, would be uncomfortable.
- The contrast between the artistic craft form of the legs and glass and the appearance of the accurately engineered fittings also contributes to the chair's aesthetic appeal.

In conclusion, the aesthetics of this chair are its most important feature, while its function as a chair is merely secondary, quite different from Modernist designs described earlier.

Etruscan chair

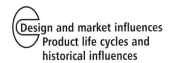
Design and market influences
Product life cycles and
historical influences

PRODUCT ANALYSIS EXERCISE: *historical influences*

Bauhaus chair

Study the picture of the B32 chair designed by Marcel Breuer. Describe how this product was influenced by Bauhaus principles. Your answer should make reference to materials, construction method, function, aesthetics, and so forth.

PRODUCT ANALYSIS EXERCISE: *historical influences*

Super Lamp

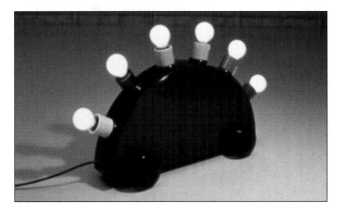

Study the picture of the Super Lamp, designed by Memphis Group designer, Martine Bedin. Explain how its design was influenced by Post-Modernist principles. Your answer should make reference to materials, construction method, function, aesthetics, etc.

Exam questions

AS exam question

1. Choose one of the design periods listed below, and answer the following questions:
   - Modernism;
   - Post Modernism;
   - Contemporary.

   (a) Name a designer from your chosen design period. [2]

   (b) Draw an example of their work and label the key design points. [10]

   (c) Explain how the design has been influenced by the particular design style of the period. You should make reference to aesthetics and the function of the product in your answer. [12]

A2 exam question

1. (a) The teapot shown was designed by Marianne Brandt at the Bauhaus in 1924.
   (i) Explain what the Bauhaus was. [4]
   (ii) What were the principles of the Bauhaus? Explain how these influenced the design of the teapot. [8]

   (b) With the aid of diagrams, describe the work of a Post Modernist designer you are familiar with and explain how Post Modernist principles influence the design. [12]

Design and nature

Introduction

Designers often use things found in the natural world to inspire new designs. For centuries, craftsmen and craftswomen have observed nature and re-created it in the form of jewellery, textiles and art. Alternatively, designers and inventors have looked towards nature in order to find solutions to design problems.

At AS level

As an AS level student, you should be aware of some common examples of where designers have used nature to inspire designs.

At A2 level

A2 students should be able to analyse some products in detail, showing a more in-depth understanding of how the products have evolved from specific things found in nature.

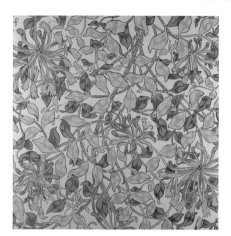

A William Morris print

Craft inspired by nature

The Arts and Crafts Movement designer William Morris (1834-96) used natural things, such as birds and plant life, and re-interpreted them into stylish wallpapers, tapestries, tiles and stained glass. Morris did not simply copy nature; he instead produced stylised designs from it. This idea became popular with later designers such as Charles Rennie Mackintosh (1868-1928) who used motifs based on natural forms, such as flowers, in many of his designs at the Glasgow School of Art.

Flight inspired by nature

The first powered flight was achieved by the Wright brothers in 1903, but people had been studying flight and how birds fly for centuries before this. It was long realised that the study of birds would one day provide the solution to powered flight.

The shape of a bird's wing creates high pressure over the one surface and low pressure over the other, which in turn helps to create lift.

A Rennie Mackintosh chair

The principle of flight

- As air hits the front of the wing, it is split by the leading edge (A, B).
- As air flows over the top of the wing, a low-pressure area is created (A).
- Air flowing under the wing travels at the same speed as that over the top (B).
- There are drag effects due to the air hitting the wing then being divided over the top and under the bottom, but the net result is that lift is created (C, D).
- The greater the speed of air passing over the wing, the more lift is created (E).

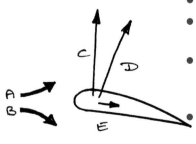

Principle of flight

Types of flight

Birds fly in different ways. Humming birds, for example; have small, narrow wings that beat very fast allowing it to hover; while birds like eagles have long, broad wings for soaring on thermals. Swallows and swifts have long narrow wings allowing them to fly extremely quickly. Tucking the wings in towards the body can increase speed even further. Peregrine falcons can reach speeds of 120 m.p.h. as they swoop for a kill.

A variety of aeroplanes have been designed to mimic these different ways of flying. Gliders, for example, have long narrow wings enabling them to produce large amounts of lift and rise quickly through thermals. Fighter aircraft have small broad wings, which effectively reduce drag and give the aircraft considerable manoeuvrability. Some fighter aircraft can increase their speed by employing 'swing-wing' technology in the same way that peregrine falcons and swallows can.

Most aircraft wings have a similar shape in cross-section to that of birds. Aircraft designers also noted that birds are able to adjust their wings and tail feathers in order to make turns. This information helped in the development of the flaps and tail rudder controls in aircraft. It could also be argued that the shape of a bird's body may have helped to influence the aerodynamic shaping of early aircraft, in order to minimise drag.

Inventions from plant life

The hook and loop fastening system (known more commonly by its trade name 'Velcro') was invented by George de Mestral in 1948. De Mestral was said to have been inspired by seed burrs that stuck to his trousers when he was walking his dog. He examined the seed burrs under a microscope, together with the fabric of his trousers. He discovered that the seed burr had hundreds of tiny hooks that stuck in the weave of the material, which made lots of loops. He later commissioned weavers to produce the hook and loop fastening system that is today used in a wide range of commercial applications.

The Catseye

The Catseye system, used to mark the centre or edge of a road, was invented by a road repairer from Yorkshire called Percy Shaw in 1933. The story is that he invented it after the reflection of his headlights from a cat's eyes saved him from going off the road on a dark and foggy night. He realised that if the 'eyes' could be manufactured, he could make a product that would help people drive more safely.

A Catseye for road use is typically made from glass with a piece of foil in the back to act as a reflector. The Catseye is housed in a rubber block, so that when a vehicle runs over it the eyes are pushed down in the block and are cleaned by the sides of the block at the same time.

Angle-poise lamp

George Carwardine was an automotive engineer specialising in suspension systems. He realised that he could use springs to act as 'muscles' to balance an incandescent light bulb in space. He commissioned Herbert Terry & Sons of Redditch to manufacture the springs. He designed two lamps: one for industrial use and one for domestic use.

The springs hold the lamp in any desired position within its range of movement.

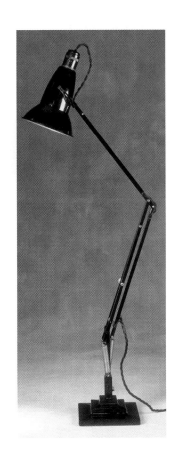

Table 19: Construction of an angle-poise lamp

	Material	Construction
Lamp housing and base	Mild steel	Spinning and piercing
Arms	Mild steel	Cut to length from stock square, hollow, section tube Pierced (for fastenings to 'elbow' pivot)
Support bracket	Mild steel	Blanked from sheet material Press formed and pierced
'Elbow' pivot	ABS	Injection moulded

Task

(a) State why mild steel is a suitable material for the lamp housing.
(b) Suggest a suitable finish for the lamp components and state why that finish is appropriate.

The 'elbow'

The angle-poise lamp has an 'elbow' pivot, which allows it to straighten and bend in much the same way as a human arm – giving approximately the same range.

The tension springs on the lamp work in the same way as the biceps and triceps of the upper arm.

- When the lamp is lowered, or bent, one set of springs is relaxed, while the other is in tension.

- In the same manner, when the forearm is bent the biceps contract and the triceps relax.

Task

Sketch and describe two products where nature has influenced the design.

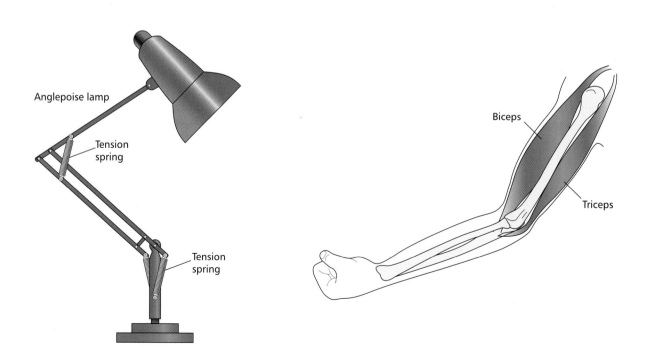

Anglepoise lamp

Tension spring

Tension spring

Biceps

Triceps

PRODUCT ANALYSIS EXERCISE: *design and nature*

Monocoque car body shell

Modern car bodies are made by joining press-formed sheet steel together. The structure that is made is called 'monocoque'.

1. Explain why mild steel is an appropriate material for the manufacture of car bodies.

2. Use notes and diagrams to explain how the various press-formed components are joined together.

3. Use notes and diagrams to explain how nature may have influenced the design of monocoque body structures.

4. What are the benefits to (a) the manufacturer, and (b) the consumer of producing car bodies in this way?

Exam questions

AS exam question

1. Parasols, used in garden furniture to shade the user from the Sun's rays, are an example of a product influenced by nature.
   (a) State the type of structure represented by the parasol.   [2]
   (b) In what way has nature influenced the design?   [10]

A2 exam question

1. Nature provides numerous examples of structures that are useful to designers. Describe **three** different examples of how nature has influenced the design of products.   [3 × 8]

Glossary

A

Adhesive Chemicals used to bond materials together. Different adhesives can be used to bond specific materials.

Aesthetics The features in a product that make it visually appealing, e.g. colour, texture, shaping, styling features.

Alloy A mixture of two or more metals with the aim of enhancing particular properties.

Aluminium A soft, lightweight, silvery-looking material. Usually alloyed with other metals to increase properties like strength and ductility. (See also Duralumin.)

Annealed glass Glass is cooled very slowly in an oven called a 'lehr' to reduce internal stresses.

Annealing A process of heating a metal that has been work hardened. 'Soaking' the metal at an appropriate annealing temperature allows the crystals to reshape, so relieving internal stresses that cause work hardening. Annealing is followed by a very slow cooling.

Anodising An electrochemical process that is used to make the surface of aluminium more durable. In addition, coloured dyes can be added for a decorative finish.

Anthropometrics The use of body measurements to determine the optimum size for products for comfortable and efficient use.

Applied finish Refers to coatings that are applied to the surface of the material for protection and/or decoration.

Artificial intelligence (AI) New robotics technology aimed at developing robots that can inter-react with other robots and react to changes in their immediate environment. Such robots are able to 'think' for themselves and change their programming to adapt to changes around them.

Automatic guided vehicle (AGV) A robotic vehicle used to ferry materials or parts around a factory.

Automatic storage and retrieval systems (ASRS) These are storage racks organised in horizontal or vertical patterns. They are accessed by robots that can operate on the x-, y- and z-axes of the storage area. These robots are usually mounted on a gantry that runs under the storage racks. The robot can place or retrieve pallets and load them onto AGVs.

B

Bauxite The most common metal in the Earth's crust – can be made into aluminium using processes such as electrolysis.

Biodegradable A term given to materials that will break down with the aid of natural processes such as sunlight and rain.

Biopol A polymer made from natural cellulose, which can be used to make biodegradable packaging.

Biscuit The name given to the ware, once it has been fired and before any glaze or decoration has been applied.

Block models Concept models made from materials like MDF, clay and jelutong, to prototype product designs. Block models can be used to give clients an accurate impression of a product, to test ergonomics and to take dimensions from for manufacturing moulds, etc.

Blowing Used to produce an air bubble in the first gather of glass, then to enlarge the 'bubble' to produce the final object. Traditionally glass is blown by a glass blower, but some manufacturing processes, e.g. bottle manufacture, include blowing the material into a mould.

Blow moulding A process of manufacturing thermoplastic materials into re-entrant shapes with a single opening. Examples include soft drinks bottles, detergent bottles, etc.

Blueing A method of finishing steel products that involves heating the product to around 300 °C followed by quenching in oil. It is the oil that gives the material a characteristic 'blue' finish.

Boro-silicate glass Glass containing boron and used generally for kitchenware and laboratory equipment, due to its heat resistance and good resistance to chemicals.

Brazing A form of joining using heat and a brazing spelter filler rod. Useful for producing a strong joint in steels and copper.

British Standards Insitution (BSI) An organisation dedicated to producing British (BS) and European (EN) quality and safety standards, and testing products against those standards for companies wishing to register their products.

Burr A rough edge following a cutting process.

C

CAD models The production of computer-generated drawings using vector or raster graphics. CAD models can be simple dimensioned drawings or, at the other extreme, animated 3D simulations.

Calendering A process used to manufacture thermoplastic sheet. Generally PolyEthylene (PE) and PolyVinyl Chloride (PVC) are used to create the sheet.

Carbon Carbon is an element found in a range of materials. Carbon is found in most organic materials – materials like coal and diamond are made up purely of carbon. The amount of carbon in a steel has the effect of increasing strength and hardness. (See also Heat treatment.)

Catalyst A substance that increases the rate of a chemical reaction without being changed itself by that reaction. For example, hardeners used with resins in the production of glass-reinforced plastic (GRP) products.

Cathode ray tube (CRT) Still used in most television sets for displaying a transmitted image and in computer monitors for displaying data or graphics. The large face of the tube is the screen of the television or monitor.

Cellular manufacturing Refers to organising production into cells of machines performing different tasks. They are typically laid out in a U-shape, rather than a production line, to enable one person to operate several machines.

Cellulose A constituent part of timber. Approximately 55% of a tree is made up of cellulose. *Note*: This material can be used to produce a cellulose-based polymer.

Cermet A mixture of a metal and a ceramic.

ChloroFlouroCarbons (CFCs) Gases used in refrigerators, and in some plastics, to make foams, etc. CFCs have been linked to damaging the ozone layer. The ozone layer prevents harmful radiation from the Sun reaching the Earth's surface.

Computer Numerical Control (CNC) The control of machines such as lathes, drills, punches and so on, using a special machine programming language, which can be generated from CAD drawings.

Colour Liquid Crystal Displays (LCD) Liquid crystals are carbon-based compounds that can be made to move in response to a small voltage. When in their 'natural state', they allow light to pass through. When charged, the crystals align with the flow of electrons, making a pattern or block of colour.

Compaction The act of forcing powder particles together. Cermets are formed in this way. Compaction takes place between dies and provides the form of the product prior to sintering.

Compression moulding The manufacturing process generally used for processing thermosetting polymers.

Computer-aided design (CAD) The use of software that can convert CAD drawings into CNC machining data.

Computer-aided engineering (CAE) The use of computers to model engineering problems and simulate working conditions to see how they perform.

Computer-aided manufacture (CAM) The use of computer numerically controlled (CNC) machines to increase efficiency of production.

Computer-integrated manufacture (CIM) The use of computers to link together business and manufacturing data and the control of production, in order to make production more efficient.

Computer simulation The use of computers to try out manufacturing processes or test product functions, etc.

Control of Substances Hazardous to Health (COSHH) Regulations dealing with the safe handling, use and storage of hazardous materials.

Conversion Sawing up logs to provide usable wood forms.

Co-polymerisation The combining of two or more monomers (separate groups of molecules of a material) to create a new material. The new material is then called a co-polymer.

Copper A brownish-looking metal; can be alloyed with zinc to produce brass, or with tin to produce bronze.

Copy Text-based information.

Copyright The legal right of ownership of copy or artwork that cannot be copied for use without the owner's permission.

Corrosion The deterioration of metals.

Cradle to grave This refers to assessing the life cycle of a product and its impact on the environment.

Crimping A joining method used in thin sheet steel to join two shapes together. For example, the top of a soft drinks can.

Crop marks These indicate where a sheet of printed material is to be cut.

Cross-link The fixing together of the long-chain molecules within a polymer.

Crystal The main building block of most, if not all, metals.

Cullet Glass that has been crushed into very small particles ready for re-melting and recycling.

Cupping A form of warping due to a combination of conversion and uneven seasoning.

Curing Forming rigid cross-links in thermosetting materials.

Cycle time The time it takes to complete an operation within a production process.

D

Deathwatch beetle Insect responsible for the destruction of mainly hardwoods, e.g. oak-frame buildings; churches and barns can be affected by the deathwatch beetle.

Deburring The process of removing burrs from the edge of material that has been cut.

Decay The deterioration of woods. (See Dry and Wet rot.)

Deciduous A general term used to describe trees that lose their leaves in autumn.

Degradation The deterioration of polymers.

Degrees of freedom This refers to the number of joints a robot arm has and each direction a robot or part of it can move, determining how flexible the robot is. The more degrees of freedom, the more useful a robot is, as it can operate in confined spaces.

Demand pull Consumer demand can lead to developments in products, e.g. demand for more energy-efficient domestic appliances as fuel costs rise and consumers are more aware of green issues.

Desktop publishing (DTP) The use of computer software to arrange images and text on a page for printing purposes. The data produced can be downloaded directly to a digital printing system.

Digital printing Printing presses can be linked directly to computers to remove the need to make printing plates.

Dry rot Is carried by fungal attack causing a breakdown of lignin resins that hold the cellular structure together. The fungus is *Merulius lacrymans* and is usually present in dry, unventilated areas. The strands of the fungus can also penetrate brickwork, enabling it to travel quickly through a building.

Duralumin The general name given to alloys containing aluminium.

E

Electrochemical cell The term given to the conditions necessary to promote corrosion between two different metals. The additional component is an electrolyte. An example of an electrolyte is rainwater containing salts and acids.

Electronic data interchange/exchange (EDI/EDE) The use of computers to transfer information, e.g. transfer of CAD data to a manufacturing database.

Electronic point of sale (EPOS) The use of barcode readers to send sales data to distributors and manufacturers in order to maintain correct stock levels.

Electronic product definition (EPD) This involves using CAD/CAM software, where all the product designing and manufacturing information is processed and stored on a central database. All of the production team have access to the database and it is updated as the product is developed.

Electroplating The use of the process of electrolysis to coat a base metal with a second, more decorative metal. For example, silver-plated cutlery.

Electrostatic spraying An electrostatic charge is set up between the object to be painted and the paint particles, making a secure bond between paint material and product surface.

End effector Robot arms are terminated with an end effector. This is a gripping hand, paint sprayer, welder or other device for moving objects around.

Epoxy resin A thermoplastic material made up of two parts – resin and hardener.

Ergonomics The study of the interaction between the human body, products and environments.

Evergreen A general term used to describe trees that do not lose their leaves in autumn.

Exploded views Drawings showing construction of a joint or assembly details of a product or part.

Extrusion Used to produce plastic products that have a uniform cross-section. Examples include curtain rails, window frame sections, guttering, etc. The process can also be used to insulate wires and cables with a polymer.

F

Fastening Any product that is designed to hold materials together. Examples include spring clips, self-tapping screws, and nuts and bolts.

Ferrous alloys A mixture of two or more metals – at least one of which contains iron (ferrite) and carbon.

Ferrous metals Metals that contain iron (ferrite) and carbon.

Fibre-based composite A material that is made up of resins and fibres. Can also refer to materials such as reinforced concretes, where reinforcing rods (fibres) have been added to the mix.

Fibre optics Glass fibres that are used to transmit data using pulses of light.

Fibre-reinforced polymers (FRP) For example, carbon-fibre reinforced polymer, Kevlar-reinforced polymer, glass-reinforced plastic. These are composites that are made by combining a woven material, such as carbon fibre, with a polymer resin and a catalyst to make a strong, lightweight material.

Filler rod A material that is used to help create the joint between two materials. For example, a solder filler rod is used when joining copper pipe together.

Fine bone china A form of clay that contains ground-up animal bone. It has a translucent quality, which means that some light will pass through the material. Fine bone china is used in the manufacture of products aimed at the upper end of the market, by producers such as Wedgwood and Royal Doulton.

Finishing Refers to the removal of burrs or other blemishes in a material following processing.

First generation Basic robots with limited sensing; normally used for loading, transfer of components, etc.

Flexible manufacturing systems (FMS) Organising production equipment to allow manufacture of a variety of different products, as opposed to dedicated systems that can only make a single type. Such production allows for changes in consumer demand.

Float glass Sheet glass is produced by floating molten glass on a bath of molten tin. The flat surface of the molten metal gives the glass its flat, smooth surface.

Flux A chemical used to prevent oxidation of the material at the joint area just prior to joining. Borax is used as a flux when joining by brazing. It is known as an 'active flux', meaning when heated it will clean the joint as well as keep it clean during the joining process.

Function How a product satisfies the intended purpose.

Fused deposition modelling (FDM) A type of rapid prototyping machine that extrudes layers of liquid polymer to build up a model.

G

G-Code A programming language to control the movement of tooling on CNC machines.

Glaze A slurry of finely ground glass particles. Colours may also be added to produce coloured glazes. Glaze is used to coat the ceramic ware in its biscuit form and can be applied by dipping or by spraying.

Global manufacturing Modern-day industrial practice of designing in one part of the world and manufacturing the product in another where materials and labour costs may be cheaper.

Global warming A layer of greenhouse gases in the upper atmosphere trap heat from the Sun and prevent it reflecting back into the outer atmosphere and beyond. This is said to be raising the temperature of our atmosphere and may be leading to climate change.

Grain The visual effect of the flow of tracheids.

Graphical user interface (GUI) What you see on a computer screen. It is made up of the icons, toolbar buttons and drop-down menus available that make it possible to activate and react to events on a computer screen via a keyboard or mouse, etc.

Greenhouse gases Gases such as carbon dioxide (CO_2) produced as a result of burning fossil fuels. These have been linked to global warming.

H

Hardening Hardening of steels is carried out by heating the metal (usually steel) to cherry red followed by rapid cooling (quenching). This process is usually followed by tempering to remove any brittleness.

Hardening and tempering These two heat treatments are generally thought of as being carried out in sequence on a metal – generally steel.

Hard soldering Hard solders, sometimes called 'silver solders', are a range of solders that have different melting points enabling complex assemblies to be completed in steel, copper or brass.

Hardwood Timbers that are deciduous and slow growing.

Health and Safety Executive (HSE) A government advisory service that helps companies meet health and safety obligations under the Health and Safety at Work Act. The HSE publishes safety posters, books and copies of specific Health and Safety regulations. Local HSE officers visit employees to check they are complying with regulations and to investigate accidents.

Heat source Any piece of equipment that applies heat in an appropriate manner for the purposes of joining materials. For example, oxy-acetylene equipment is used as a heat source for gas welding and brazing.

Heat treatment A term given to a range of processes using heat to cause a change in a metal's properties by making changes to the internal structure of the material. Annealing, hardening, tempering and normalising are the more common heat treatments.

High definition television (HDTV) A form of television broadcasting made possible by the advent of digital

TV. HDTV supports a number of formats including those suitable for large TV screens.

I

Injection moulding A method of processing thermoplastic materials. Products generally have complex 3D shapes.

Integrated circuit (IC) Also known as a 'chip': a small electronic device made out of a chip of semiconductor material. These devices contain electronic circuits in which there may be thousands of transistors.

Ion A charged atom formed when, for example, copper comes into contact with oxygen in the air. Two free electrons in a copper atom move over to an oxygen atom. This makes the copper atom a positively charged ion while the oxygen atom becomes a negatively charged ion.

Iron Iron (ferrite) is converted from its ore by heating. The resulting impurities (slag) are removed from the furnace leaving a soft greyish metal once cooled. Iron is rarely used without combining with carbon – this gives it greater strength. The result of this combination is steel. This can be alloyed with other metals in order to enhance particular properties.

Isometric drawing 3D drawing using 30° lines; used to create design idea sketches.

J

Just in sequence A refinement of JIT, in which parts are not only delivered at the right time and right place but also in sequence to match the flow of a product through an assembly line. This totally eliminates the need for storing stock at the side of the line.

Just in time (JIT) The organisation of production so that customers get their orders just in time. This avoids carrying stock of materials and components, and storing finished goods.

K

KD fittings Knock-down fittings, which come in a variety of guises, each type intended for a particular application. Used extensively (but not exclusively) with self-assembly furniture.

Kerning Done to make detailed and often small changes to the spaces between letters, increasing or decreasing the width.

Kiln drying A form of seasoning that uses steam in a controlled way to reduce the content of moisture in timber.

Knots Natural defects found in timber – the start of branches from the trunk.

L

Laminate Building up layers of materials, e.g. plywood; also refers to a layer of plastic material, e.g. melamine formaldehyde over chipboard.

Layered object modelling (LOM) A type of rapid prototyping machine that cuts layers of self-adhesive card or paper, which are then assembled into a 3D model.

Laying up The act of laminating, for example, glass fibre matt and coating it with polyester resin.

Layout The way text and images are arranged in a specific space. It includes position, size, orientation and how they work together.

Lead glass When lead is added to glass, it improves clarity and the ability to reflect light (refraction). Products include optical products such as prisms and lenses. Expensive 'crystal' tableware is made with this material.

Lead time The time a customer must wait to receive a product after placing an order.

Leading Used by typographers for a space between lines of type, increasing or decreasing its height. (In early printing processes, lead strips were used to create these spaces.)

Lehr A furnace in which hot glass (about 500°C) is placed after working, enabling it to be brought down to room temperature very slowly.

Light-emitting diode (LED) A semiconductor made up of a single junction of n- and p-type materials. as with all semiconductors current will flow when the junction voltage has been reached but in this case light is emitted. Various colours of LEDs can be used.

Lignin The natural resins that hold the cells together in timbers.

Lime-soda glass Used for windowpanes, storage jars and bottles.

Line balance This is achieved when all of the operations on a production line take the same amount of time, so that none of the operators are kept waiting for their turn to work on a component. This is done by adjusting the task done by each employee, re-organising the layout of the line, or by finding ways to speed up the slowest tasks.

Line bending A method of processing sheet materials, in particular thermoplastics; limited to producing simple shapes.

Liquid crystals Carbon-based crystals that can have their orientation changed when an electric current is passed through them.

Long-chain molecule The main constituent part of a polymer. The long-chain molecule is made up of a series of atoms – in the case of the hydrocarbons these would be made up of hydrogen and carbon along with elements like oxygen.

M

Master production schedule (MPS) A scheduling system used to organise the work to be completed within a set time period.

Materials requirement planning (MRP) A software system for work cells to order materials and components from their suppliers, when they require them.

Matrix A matrix material holds all of the fibres together in a fibre-based composite and fills in any gaps between fibres. Typical matrix materials include thermosetting polymers, such as polyesters and epoxy and phenolic resins. PEEK (a thermoplastic polymer) may also be used as a matrix, as can nylon and polyethylene.

Matt Loosely woven fibres that make up the material once resins have been added.

M-Code A programming language to control the speeds of machines, tool changing, coolant and other machine functions.

MIG welding (metal inert gas) A form of resistance

welding, similar to electric arc welding. In the case of MIG welding, an inert gas is used to provide a protective cloud around the joint area as it is being joined.

Mock ups Rough models, often full-size, made from low-cost materials, such as card, MDF, etc.

Moire effect Blurring of an image caused by inaccurate registration.

Moisture content The amount of 'water' in the timber. Usually shown as a percentage of volume.

Mood boards Collection of images, colours, fabric/material samples, etc. used to guide designs towards a style.

N

Nail Used for joining woods; not appropriate for man-made boards. Available in a variety of forms, such as oval and round wire nails, masonry nails and panel pins. Special forms are used for upholstery and fixing roofing felt.

Natural barriers Protective layers close to the surface of a material that protect the material from corrosion, decay or degradation. An example is the oxide layer that is present on non-corrosive metals, such as stainless steel or copper.

Non-ferrous alloys Mixtures of two or more metals – none of which contain iron (ferrite).

Non-ferrous metals Metals that do not contain iron (ferrite).

Normalising A heat treatment that is applied to steels in order to obtain crystals of a smaller, more regular size, thereby making the material stronger and tougher.

N-type A type of semiconductor that has been doped with a more negatively charged material allowing a flow of energy in the same direction as the current flow in a wire.

O

Obsolescence Products become obsolete as they are superseded by better models. Some products are said to have 'built-in' obsolescence for safety, or to ensure demand for new products.

Occupational therapy (OP) Hospital department that helps patients learn how to use home aids or adapted equipment and to cope with living with a disability.

Off-line programming A method of programming a robot using a virtual reality software system.

One-piece flow A system where products proceed in a production line one at a time, without interruptions or waiting time.

On-glaze pattern A decorative pattern that is applied to the ware after it has been glazed.

Orthographic drawings Drawings showing the front, plan and end views of a design. These may include dimensions and details of materials.

Oxidation When a material comes into contact with oxygen (e.g. in the air) the result is an oxide layer that forms over the surface of the metal. In most metals this serves to protect the material from further oxidation, but in the case of steels this oxide layer is porous and so allows further contact with air and so further oxidation.

P

Pad printing A method of transferring the inks that have been deposited by screen printing onto the product to be decorated. The pad is heated so that it will lift the inks more readily.

Painting The process of applying paints, e.g. brush painting, spray painting. The term applies to all types of paint including oil-based gloss paints, acrylic paints, etc.

Pantone colour A specific colour recognised by a code and generally used as a spot colour.

Patterned glass Patterned, or frosted, glass is produced by passing softened glass between two (cast iron or stainless steel) rollers. One of the rollers contains the pattern, while the other is smooth.

Parison The extruded tube of thermoplastic material used in the process of blow moulding.

Particle-based composite Composites that consist solely of particles of two or more different materials, e.g. cermets and concrete.

50th percentile The average or most common anthropometric measurements from a sample.

95th percentile The upper limiting anthropometric measurement. Designers usually produce designs suitable for body sizes between the 5th and 95th percentile as this takes in the majority of users.

Performance How well a product carries out its function.

Permanent A joining method where separation of the component parts results in damage to the materials. Examples include a welded joint.

Perspective drawings 3D drawings using lines that converge towards either a single vanishing point or two vanishing points. These give drawings a more realistic appearance and can be used to view designs from a variety of angles.

Phosphorescent pigments Ceramic powders that have the ability to absorb light and then release the light energy over a long period of time. Can be mixed with acrylic paints/inks for creating illuminated signs that do not require an additional power source.

Piezoelectric devices Devices that either generate electricity when loaded, or change shape/size with a useful force when connected to an electronic circuit.

Plastic coating The process of heating a metal (usually mild steel) product to around 230 °C and then dipping it into a fluidised bath of thermoplastic plastic granules, which stick to the product. A smooth plastic finish is achieved.

Plastic dip coating A method of finishing a metal-based component. The component is heated to above the softening point of the polymer – which is in the form of a fine powder – then dipped into the polymer. The polymer adheres to the heated metal, which begins to cool, and so solidifies into a protective layer.

Plate glass A high-quality glass of few impurities that has been rolled and polished. Uses include mirrors and large windows.

Points and picas Special units of measurement used for type.

Polymer The proper term for a plastic material.

Powder coating This process uses an electrostatic charge to coat the metal product. Once coated, the product is baked in an oven to produce a smooth, high-quality finish.

Powder pressing The press forming of clay powder using hydraulic presses. Typical products include dishes, plates and saucers. Also known as 'dust pressing'.

Powder processing The range of manufacturing processes for metal or ceramic powders.

Presentation boards Boards usually used to present designs to a client or others. These may have rendered drawings, dimensioned orthographic views, and so on.

Print run The number of copies to be printed from an original at any one time.

Process colours Cyan, magenta, yellow and black – the four colours combined in a print process to make full-colour images.

Product data management software (PDM) This is software used to bring together all of the computer systems used, including CAD/CAM and CIM. This software enables: designs to be simulated on screen; production and business information to be stored accurately in a database; production to be monitored. PDM software enables products to be designed in one part of the world and manufactured in another, as designs, materials specifications and even CNC programmes can be communicated electronically.

Proof A realistic sample of a page layout, or other printed item, used for inspection prior to final printing – a printer's final prototype.

P-type A type of semiconductor that has been doped with a more positively charged material allowing a flow of energy in the opposite direction to the current flow in a wire.

Pyrex A well-known heat-resistant glass containing boron. (See Boro-silicate glass.)

Q

Quarter-sawn A form of conversion that can prevent warping and can be used to enhance the grain.

Quenching The term given to the rapid cooling of a metal following a heat treatment.

Quenching media The material used to achieve quenching – brine, water, oil, air, etc.

Quick response manufacturing (QRM) The organisation of production to manufacture to customer demand, rather than manufacturing items to stock.

R

Rapid prototyping (RPT) The use of CNC machines that create 3D objects using lasers to solidify liquid polymers, known as stereo-lithography.

Registration marks Marks printed with each colour in a two-, three- or four-colour printing process that should line up to create a sharp clear image.

Renderings Colour drawings using tones of colour and texture to produce a realistic artist impression of designs. These can be done using CAD.

Renewables Materials that are extracted from managed sources, such as Scandinavian pine taken from forests where trees are replaced by saplings as they are felled.

Renewable energy Energy produced from sources such as wind, water and solar power. These are referred to as 'infinite', because they will never run out.

Resistance welding The joining together of metals with a heat source generated by electric current. Electric arc welding, seam welding and spot welding are examples of resistance welding.

Rigid cross-link These types of cross-links are found in thermosetting plastics and elastomers. They are a more rigid bond than the van der Waals bond found in thermoplastics, and do not react to the application of heat.

Risk assessment A document assessing the type of hazard, the level of risk, who might be affected by the hazard, and a description of control measures taken to minimise the risk associated with using specific materials and manufacturing processes.

Rolling The process of rolling clay onto a mould in the manufacture of, for example, dinner plates.

Rot The breaking down of the lignin resins in woods that hold the tracheids (cells) together.

Rotational moulding A method of moulding thermoplastic polymers that involves plastics in powder form taking the shape of a fully enclosed mould as the mould passes through a heating section. The material becomes rigid as the mould is cooled.

Rust The oxidation of steels (not stainless steel) due to contact with the air that may well contain moisture. The oxide layer of steels is porous so further oxidation can take place. This results in a thicker oxide layer as the material breaks down.

S

Scale models Used to test designs without the expense of making full-size prototypes; can be used to test ergonomics, construction methods, colour schemes, and so on.

Seam welding A form of resistance welding used for joining seams in fabricated tubes and in food cans.

Seasoning Drying timber to a known moisture content.

Second generation Robots that use sensors to detect faults, changes in their immediate environment and diagnose problems.

Self-coloured The term given to a material that has an acceptable colour after processing. Plastics are said to be self-coloured when they have their colour added prior to processing and, therefore, appear as the desired colour.

Self-finishing The term given to a material that has an acceptable high-quality finish after processing. Generally applied to plastics that do not need any further processing to produce the required finish; they can be given a very high-quality finish by being produced in moulds that have a high-quality finish. The term can also be applied to: metals, such as stainless steel, copper, brass; woods, such as oak and teak which, depending on application, may not require further processing to produce the required finish.

Self-tapping screws These screws are used for forming their own threads in materials. They can be used to join metals and plastics and can also be used with various forms of captive nut, for example, in the assembly of car interiors.

Semiconductor A material that will conduct electricity only under special circumstances, e.g. when it has reached a specific temperature or when a particular voltage has been applied. By doping the material with small amounts of 'impurities' (i.e. other elements) the temperature and/or voltage can be more closely controlled.

Shape memory alloys Metal alloys specially developed to 'remember' their shape under specific conditions.

Glossary

Shrinkage All timbers shrink due to moisture lost in seasoning.

Silica Sand – one of the raw materials of glass.

Silkscreen printing A method of applying colour in the form of a pattern to a glass substrate, prior to pad printing onto ceramic ware. The pattern is formed on a mesh (screen), and a single colour is forced through the screen by a squeegee onto a glass substrate. One colour can be printed at a time with this method. Having up to five stations will mean up to five colours can be applied to the ware.

Sintering A process whereby powder particles are fused together at their contact points between other particles.

Slab sawn A form of conversion where the trunk of the tree is cut into slabs – more prone to warping.

Slip Liquid clay – clay mixed with water.

Slip casting The process of pouring slip into plaster of Paris moulds.

Slumping A process where sheet glass is heated to a temperature where it softens, and is allowed to take the shape of the mould.

Smog A layer of pollution gases created from traffic exhaust fumes, often seen in densely populated cities on hot days and causing breathing difficulties in some people.

Softwood Timbers that are evergreen and quick growing.

Soldering A method of joining using heat and a low melting point filler of solder. This technique is used to join copper pipe for domestic water and heating systems.

Solid models These show a full 3D image and are used to show what products would look like when manufactured. They contain full data for machining and for internal details of products. As they contain lots of data, they take time to process and regenerate, which in turn requires powerful computers.

Solvent Any chemical that will dissolve a material. For example, a thermoplastic produced from crude oil will begin to deteriorate in the presence of oil or its vapour.

Splits Separation of timber fibres along the grain due to rate of drying.

Spot welding A form of resistance welding where the metal components being joined are clamped between two electrodes. An electric current is passed between the electrodes causing a build-up of heat at the centre of the joint, fusing the two components together. Used extensively in the assembly of car body panels.

Steel Steel is a mixture of iron and carbon. The amount of carbon present in the material determines the material's initial strength. Alloying with other metals enables specific properties like strength, toughness and resistance to corrosion to be enhanced.

Surface models CAD drawings that provide a more realistic 3D image than wire frame. They can be used to show machining tool paths and data, and are quicker to re-process than solid models.

Surface mount components Sometimes known as SMT (surface mount technology); refers to tiny electronic components, such as resistors and capacitors, soldered to the surface of circuit boards. These are a fraction of the size of traditional 'through-leg' components, which have to go through a circuit board and are soldered on the opposite side.

Swatch Sample showing colour, texture, etc. (e.g. of fabric, paper), used as a portable source to select from.

T

Takt time The time available to make a product if production is to meet demand. The term 'takt' is from the German and means 'a musical tempo or beat'.

Tanalising A process whereby wood is pressure-treated with preservative.

Teach pendant A method of programming a robot by using a hand-held remote control device or key pad.

Technology push New products can be developed and old products made obsolete by advancements in technology, e.g. smaller, more powerful microprocessors led to the development of computers with improved performance.

Telematics A system for tracking a product from customer order through to manufacture and dispatch.

Tempered glass Another term for toughened glass.

Tempering A heat treatment process carried out after hardening to remove any brittleness that may be present in the hardened material. The hardened material is cleaned to its natural shiny grey colour. Heat is then applied and when the correct tempering colour is seen the material is then rapidly cooled (quenched).

Tempering colour The colour seen on a piece of steel that will indicate an appropriate temperature to remove brittleness from the material.

Temporary A temporary joining method is one where the components can be joined and separated without damage to the material. An example would be a nut and bolt assembly.

Texture The 'feel' of the surface of the material; can range from extremely smooth, through patterned, to very rough, e.g. for grip.

Thermochromic pigment A smart material that changes colour in response to changes in temperature. It can be used in products to indicate, for example, whether food or drink that it contains is too hot to consume.

Thermoforming A form of vacuum forming. The main differences are that air pressure and an additional female mould assist the forming process, to enable greater detail to be achieved.

Thin-film transistor (TFT) A TFT screen is basically a liquid crystal display that has a transistor for each pixel. This allows for rapid responses on-screen to inputs, for example from the mouse of a computer. Laptops and notebooks have TFT screens and they are becoming popular for desktop computers.

Third generation Advanced robots that use sophisticated sensing to detect changes to their immediate environment, diagnose faults and, if possible, rectify them.

Thumbnail sketches Rough sketches of a design idea.

TIG (tungsten inert gas) A form of electric arc welding using a tungsten electrode, the arc from which is surrounded by an inert gas such as argon to prevent oxidation during welding. Materials welded in this way include aluminium and stainless steel.

Total transfer printing The name given to the process of applying decoration to ceramic ware where up to four or five colours may be used. This process uses a combination of screen printing and pad printing to apply the colours to the ware.

Toughened glass Glass is heated uniformly and then rapidly cooled by air jets. This causes the outside surfaces to be under compressive stress, while the inside of the material is in tension. External forces must overcome these stresses to shatter the glass.

Tracheids The cells of woods.

Twisting A form of warping that is due to a combination of a method of conversion (sawing the trunk of the tree) and uneven seasoning.

Typography The study and design of lettering styles.

U

Ultrasonic welding The use of very high frequency vibrations to generate heat within the area to be joined, thereby allowing the materials to fuse together. Plastics and some metals can be joined in this way.

Under glaze pattern A decorative pattern that is applied to the biscuit ware before the ware has been glazed.

V

Vacuum forming A process where sheets of thermoplastic polymer are heated and formed to shape by the application of a vacuum to 'pull' the material down onto the simple mould.

van der Waals bond A type of atomic bonding found in thermoplastics. It is this electrostatic bond that allows the reshaping of thermoplastics when heated.

Varnish A hard transparent finish that is applied to wood. Yacht varnish is applied to products that are intended for external use; while synthetic varnishes, such as polyurethane varnish, are used internally.

Veneer A thin section sheet of timber (usually hardwood) that is glued to a cheaper base material, e.g. chipboard or block board.

Video conferencing A PC desktop system that enables designers to talk to and see manufacturers, clients, etc., while simultaneously working on CAD drawings or other tasks on the computer. Video conferencing reduces the need for key personnel to travel to meetings.

Virtual reality (VR) The use of 3D simulation software that enables designers to produce photo-realistic images of products in lifelike settings and to interact with them.

W

Walkthrough programming A method of programming a robot by physically taking a robot through a series of movements.

Warping Deforming in timber due to uneven drying.

Web-fed Printing onto a continuous roll of paper, rather than onto individual sheets (sheet-fed).

Welding The general term given to joining primarily metals, and some polymers, by heat fusing the component materials together. Polymers can be 'welded' in one of three ways:

- with the use of heat to soften the components being joined, followed by the inclusion of a filler material;
- with the use of solvents to chemically melt the polymer at the edges being joined;
- with the use of ultrasound – very high frequency vibrations that agitate the atoms and molecules so they become heated. The material then softens aiding joining.

Wet rot Decay in woods that is a direct result of alternating cycles of the timber being wet followed by drying, i.e. an accumulation of moisture that breaks down the cellular structure of the timber.

Whole life cycle The cycle of a product's life from raw materials extraction through to final disposal.

Wire frame model CAD drawings of products using a range of lines, arcs and points. They take up very little memory and, in wire frame, image-processing time is kept to a minimum. However, as there are no visible surfaces, they do not show any surface or solid properties. They can also be difficult to understand.

Wood joints The general name given to the more traditional forms of joining woods. Examples include mortise and tenon joint, dovetail joint, finger/comb joint, bridle joint, etc.

Wood preservative Applied to timbers that are exposed to general weather conditions and which do not have any other means of protection. Preservatives prevent moisture entering the structure of the material, thereby reducing the risk of wet rot.

Wood products Timber-based products made from the by-products of timber.

Wood screw A form of screw designed for use specifically with woods. The two materials are held together by the shape of the screw thread that the screw forms in the material as it is being inserted.

Woodworm An insect that attacks both hardwood and softwood by laying eggs in crevices in the material. The larvae eat away at the timber then emerge through flight holes – the tell-tale signs of woodworm infestation.

Work envelope The maximum area a robot can extend to and do useful work.

Work hardening The name given to the effects of processing, i.e. rolling, bending or hammering a material while it is cold (i.e. no heat applied). Work hardening occurs when a metal is processed without the use of heat. While being cold worked, the metal's crystal structure is distorted. This in turn creates internal stresses that are responsible for making the material harder (and therefore increasing strength).

Index

Acknowledgements

Actionplus Sports Images: 31
Alamy/Elizabeth Whiting & Associates: 64, 66 (top)
Alamy/Janine Wiedel Photolibrary: 26
Alamy/Judith Collins: 184 (bottom)
Alamy/Marc Grimberg: 61 (top)
Alamy/National Trust Photolibrary: 96 (bottom right), 181
Alamy/Sciencephotos: 23 (plywood & chipboard)
Anthony Blake Photo Library/Sian Irvine: 112 (middle bottom), 163 (top)
Art Directors & Trip Photo Library: 23 (MDF), 45 (top left), 112 (top left), 136 (both), 140, 153 (top left & right)
Brian Evans: 23 (hardboard), 30 (top), 43 (all), 62 (bottom), 66 (middle), 91 (bottom), 113, 153 (bottom)
Bridgeman Art Library: v (bottom right), 68 (right), 194 (both)
Bryn Williams/Crash.net: 25 (top)
BSI Product Services: 165
Bubbles Photolibrary: 106 (bottom)
Construction Photography: 40, 57 (bottom), 91 (top), 96 (top & bottom left)
Corbis: 62 (top), 68 (left)
Corbis/Vanni Archive: 189
Corel 243 (NT): 28
Corel 257 (NT): 56 (bottom)
Corel 155 (NT): 69
Corel 768 (NT): 73 (top)
Corel 345 (NT): 106 (top right)
DaimlerChrysler: 157
Danny Lane Etruscan Chair 1986/2004 (19mm glass, polished stainless steel studding legs with stiletto heel tip feet. H880 x W400 x D500 mm. Pieces from this edition have been acquired by the National Museum, Stockholm, 1988 and the Corning Museum of Glass, New York, 1989)
Photograph courtesy of Corning Museum: 191
Denby Pottery Company Ltd: 76 (bottom right)
Design Council Slide Collection: 190 (top both), 192
DIY Photolibrary: 30 (bottom), 45 (top right), 63
Duracell: 85 (bottom)
Dyson: 100
Eyewire DT (NT): 152
Flambeau EuroPlast Ltd (www.flambeaueuro.com): 9
Heritage Image Partnership/British Museum: 193
IKEA: 59 (middle left), 65
Jaguar Daimler Heritage Trust: 177, 197

John Birdsall Photography: 175
Josiah Wedgwood Sons Limited: 73 (bottom)
Keri Stanley: 154
Letraset: 78
Luca Mazzocco - Renault F1 Team: 25 (bottom both)
Memphis Milano: 190 (bottom), 192 (bottom)
Nick Rose v (top middle left, top right and bottom left), 4 (all), 6 (left & right), 13, 16, 18, 20, 23 (blockboard & stirling board), 45 (bottom), 56 (top), 57 (top), 59 (middle right and right), 66 (bottom both), 71, 76 (bottom left), 85 (middle top), 88, 90, 95 (bottom), 98 (both), 106 (left), 112 (top right and middle top), 133, 160, 163 (bottom), 169, 182 (middle bottom)
Nokia v (bottom middle left), 6 (middle left), 182 (bottom left), 186 (right)
Photodisc 28 (NT): 12, 20 (bottom)
Photodisc 54 (NT): 87
Rex Features Ltd: 54, 59 (left), 176, 182 (bottom right)
Rimstock: 49
Russell Hobbs: 85 (top)
Science & Society Picture Library: 180, 186 (left)
Science & Society Picture Library/Science Museum: 179
Science Photo Library: 24, 38,
Science Photo Library/Hattie Young: 173
Science Photo Library/Ian Hooton: 85 (middle bottom)
Science Photo Library/Maximilian Stock Ltd: 150
Science Photo Library/Pascal Goetgheluck: 86
Science Photo Library/Philippe Psaila: 91 (middle), 95 (top)
Smithers-Oasis UK Ltd: 83
Sony Consumer Products: 182 (top)
Stockbyte 28 (NT) vi, 55
Stockbyte 30 (NT): 112
Topfoto: v (top left), 2, 110, 174, 195
Toshiba: 184 (top)
Toyota: 161, 170
TRUMPF Werkzeugmaschinen GmbH & Co. KG: 131

Cover images

Main: Pen CAD design – Photodisc 102 (NT)
Middle, left to right: Writing – Corel 481 (NT); Virtruvian Man – Corel 481 (NT); Wireframe model – DS Ergonomics; Robot Arm – Corel 682 (NT)
Bottom, left to right: Apple II computer – Owen Franken/Corbis; Apple IIvi – Apple computers; Apple iMacDV – Apple computers; Apple Mac FS display – Apple computers.

Dedication

To Sandra and Alison, for everything they've had to put up with while we were being creative.